LINEAR INTEGRATED CIRCUITS

AUTHOR

Dr. B. T. Krishna

Professor,

Jawaharlal Nehru Technological University Kakinada,

Kakinada, Andhra Pradesh, India

REVIEWER

Dr. K. Padma Priya

Professor,

Jawaharlal Nehru Technological University Kakinada,

Kakinada, Andhra Pradesh, India

KHANNA BOOK PUBLISHING CO. (P) Ltd.
4C/4344, Ansari Road, Darya Ganj,
New Delhi, India

All India Council for Technical Education
Nelson Mandela Marg, Vasant Kunj,
New Delhi, India

Dear Readers,

To prevent the piracy, this book is secured with HIGH SECURITY HOLOGRAM on the front title cover. In case you don't find the hologram on the front cover title, please write us to at contact@khannabooks.com or whatsapp us at +91-99109 09320 and avail special gift voucher for yourself.

Specimen of Hologram on front Cover title:

Moreover, there is a SPECIAL DISCOUNT COUPON for you with EVERY HOLOGRAM.

How to avail this SPECIAL DISCOUNT:

Step 1: Scratch the hologram
Step 2: Under the scratch area, your "coupon code" is available
Step 3: Logon to www.khannabooks.com
Step 4: Use your "coupon code" in the shopping cart and get your copy at a special discount
Step 5: Enjoy your reading!

ISBN: 978-93-5538-290-0
Book Code: DIP476EN

Linear Integrated Circuit
by Dr. B.T. Krishna

First Edition: 2026

Originally Published by:
All India Council for Technical Education, Ministry of Education, GOI

Exclusively Published & Distributed by:
Khanna Book Publishing Co. (P) Ltd.
4C/4344, Ansari Road, Daryaganj,
New Delhi (INDIA)

Visit us at: **www.khannabooks.com**
Write us at: **contact@khannabooks.com**

To view complete list of books, Please scan the QR Code: KPH

Printed in Bharat.

प्रो. टी. जी. सीताराम
अध्यक्ष
Prof. T. G. Sitharam
Chairman

सत्यमेव जयते

आज़ादी का अमृत महोत्सव

अखिल भारतीय तकनीकी शिक्षा परिषद्
(भारत सरकार का एक सांविधिक निकाय)
(शिक्षा मंत्रालय, भारत सरकार)
नेल्सन मंडेला मार्ग, वसंत कुंज, नई दिल्ली–110070
दूरभाष : 011–26131498
ई–मेल : chairman@aicte-india.org

ALL INDIA COUNCIL FOR TECHNICAL EDUCATION
(A STATUTORY BODY OF THE GOVT. OF INDIA)
(Ministry of Education, Govt. of India)
Nelson Mandela Marg, Vasant Kunj, New Delhi-110070
Phone : 011-26131498
E-mail : chairman@aicte-india.org

FOREWORD

Engineers are the backbone of any modern society. They are the ones responsible for the marvels as well as the improved quality of life across the world. Engineers have driven humanity towards greater heights in a more evolved and unprecedented manner.

The All India Council for Technical Education (AICTE), have spared no efforts towards the strengthening of the technical education in the country. AICTE is always committed towards promoting quality Technical Education to make India a modern developed nation emphasizing on the overall welfare of mankind.

An array of initiatives has been taken by AICTE in last decade which have been accelerated now by the National Education Policy (NEP) 2020. The implementation of NEP under the visionary leadership of Hon'ble Prime Minister of India envisages the provision for education in regional languages to all, thereby ensuring that every graduate becomes competent enough and is in a position to contribute towards the national growth and development through innovation & entrepreneurship.

One of the spheres where AICTE had been relentlessly working since past couple of years is providing high quality original technical contents at Under Graduate & Diploma level prepared and translated by eminent educators in various Indian languages to its aspirants. For students pursuing 2nd year of their Engineering education, AICTE has identified 87 books, which shall be translated into 12 Indian languages - Hindi, Tamil, Gujarati, Odia, Bengali, Kannada, Urdu, Punjabi, Telugu, Marathi, Assamese & Malayalam. In addition to the English medium, books in different Indian Languages are going to support the students to understand the concepts in their respective mother tongue.

On behalf of AICTE, I express sincere gratitude to all distinguished authors, reviewers and translators from the renowned institutions of high repute for their admirable contribution in a record span of time.

AICTE is confident that these outcomes based original contents shall help aspirants to master the subject with comprehension and greater ease.

(Prof. T. G. Sitharam)

ACKNOWLEDGEMENT

The authors are grateful to the authorities of AICTE, particularly Prof. T G Sitharam, Chairman; Dr. Abhay Jere, Vice-Chairman, Prof. Rajive Kumar, Member-Secretary, Dr. Ramesh Unnikrishnan, Advisor-II and Dr. Sunil Luthra, Director, Training and Learning Bureau for their planning to publish the books on Linear Integrated Circuits. We sincerely acknowledge the valuable contributions of the reviewer of the book Prof. R. P. Dahiya, Adjunct Professor, NSUT; former Professor, IIT Delhi, former Director, MNIT and former Vice-Chancellor, CDL University and DCR University of Science and Technology for making it students' friendly and giving a better shape in an artistic manner.

The author wish to acknowledge Kanakavalli Alivelu Mangamma, Mother, Battula Vijaya Saradhi, Brother, Battula Madhavi, Sister for their constant support. The author is grateful to the support from Reyya Satyanarayana, Brotherinlaw, Manukonda Srilantha, Sister in law, Battula Sai Rajeev, Battula Jahnavi, Reyya Chandana, Reyya Devi for their encouragement and support. Author is thankful to the teaching and non-teaching staff of the department of Electronics and Communication Engineering, University College of Engineering Kakinada, Jawaharlal Nehru Technological University Kakinada for their encouragement. Author wish to acknowledge Kukkala Aruna (Software Engineer), Dr. Yadala Gowthami, Dr. K. Rajasekhar, Dr. D. Gowtham, Sri. K. Hanumathu Krishna Prasad, Sri. K. Hari Krishna, Dr. G. Prasanna Kumar for their help in proof reading and preparation of figures.

This book is an outcome of various suggestions of AICTE members, experts and authors who shared their opinion and thought to further develop the engineering education in our country. Acknowledgements are due to the contributors and different workers in this field whose published books, review articles, papers, photographs, footnotes, references and other valuable information enriched us at the time of writing the book.

Dr. B. T. Krishna

PREFACE

At the tail end of the 20th century, there was a revolution in technology, and as a direct result of this, integrated circuits have come to play a vital part in a variety of different industries. Analogue circuits and linear integrated circuits (ICs) continue to play a significant part in modern technology, despite the fact that digital circuits are more commonly employed. Because of this, these subjects need to be discussed from the perspective of a student, and the applications of those subjects need to be presented in detail. As a result of these advancements, I became convinced that there was a demand for a textbook that covered operational amplifiers and linear integrated circuits. With the help of this book, I have made it my mission to close this knowledge gap and acquaint pupils with the design and implementation of linear integrated circuits.

This book provides information about operational amplifiers, as well as its applications and other linear integrated circuits, such as the 555 timer, voltage regulator ICs, phase-locked loop ICs, and waveform generator ICs. It places an emphasis on discussing the principles and the workings of circuits, making it easier for teachers to outline the finer aspects and making it possible for students to easily understand the concepts.

In as many places as it can, this book places an emphasis on the more practical features associated with ICs. In addition to that, it offers solutions to a considerable variety of problems. Students will be able to better prepare for standardised testing thanks to the inclusion of objective-type questions and answers to those questions. This textbook is suitable for use by students of M.Sc. (Electronics), B.Sc. (Electronics), AMIETE, AMIE (Electronics), diploma courses in electronics, and many other similar programmes because it fulfils the requirements of the curriculum that is followed by Indian institutions while teaching the topic.

Throughout the process of producing this book, I have looked to a variety of scholarly works on this topic that have been penned by authors from both India and other countries. I would like to express my appreciation to everyone who wrote and published these books. Even while every precaution has been taken to reduce the number of inaccuracies in the book, it is still possible that there are some typos and omissions. We would be happy to hear your thoughts on this matter.

Dr. B. T. Krishna

OUTCOME BASED EDUCATION

For the implementation of an outcome based education the first requirement is to develop an outcome based curriculum and incorporate an outcome based assessment in the education system. By going through outcome based assessments, evaluators will be able to evaluate whether the students have achieved the outlined standard, specific and measurable outcomes. With the proper incorporation of outcome based education there will be a definite commitment to achieve a minimum standard for all learners without giving up at any level. At the end of the programme running with the aid of outcome based education, a student will be able to arrive at the following outcomes:

Programme Outcomes (POs) are statements that describe what students are expected to know and be able to do upon graduating from the program. These relate to the skills, knowledge, analytical ability attitude and behaviour that students acquire through the program. The POs essentially indicate what the students can do from subject-wise knowledge acquired by them during the program. As such, POs define the professional profile of an engineering diploma graduate.

National Board of Accreditation (NBA) has defined the following seven POs for an Engineering diploma graduate:

PO1. Basic and Discipline specific knowledge: Apply knowledge of basic mathematics, science and engineering fundamentals and engineering specialization to solve the engineering problems.

PO2. Problem analysis: Identify and analyses well-defined engineering problems using codified standard methods.

PO3. Design/ development of solutions: Design solutions for well-defined technical problems and assist with the design of systems components or processes to meet specified needs.

PO4. Engineering Tools, Experimentation and Testing: Apply modern engineering tools and appropriate technique to conduct standard tests and measurements.

PO5. Engineering practices for society, sustainability and environment: Apply appropriate technology in context of society, sustainability, environment and ethical practices.

PO6. Project Management: Use engineering management principles individually, as a team member or a leader to manage projects and effectively communicate about well-defined engineering activities.

PO7. Life-long learning: Ability to analyse individual needs and engage in updating in the context of technological changes.

COURSE OUTCOMES

By the end of the course the students are expected to learn:

CO-1: Analyse the Differential Amplifier with Discrete components

CO-2: Come across different applications of Operational Amplifiers

CO-3: Describe the Op-Amp and internal Circuitry: 555 Timer, PLL

CO-4: Use the Op-Amp in A to D & D to A Converters.

CO-5: To Analyze different Waveform Generators

Table for co-po attainment

Course Outcomes	1-Weak Correlation, 2-Medium Correlation, 3-Strong Correlation						
CO/PO	**PO-1**	**PO-2**	**PO-3**	**PO-4**	**PO-5**	**PO-6**	**PO-7**
CO-1	3	3	2	2	2	-	-
CO-2	3	3	3	3	3	-	-
CO-3	2	3	2	3	1	-	-
CO-4	2	3	2	3	1	-	-
CO-5	3	2	3	2	3	-	-

GUIDELINES FOR TEACHERS

To implement Outcome Based Education (OBE) knowledge level and skill set of the students should be enhanced. Teachers should take a major responsibility for the proper implementation of OBE. Some of the responsibilities (not limited to) for the teachers in OBE system may be as follows:

- Within reasonable constraint, they should manoeuvre time to the best advantage of all students.
- They should assess the students only upon certain defined criterion without considering any other potential ineligibility to discriminate them.
- They should try to grow the learning abilities of the students to a certain level before they leave the institute.
- They should try to ensure that all the students are equipped with the quality knowledge as well as competence after they finish their education.
- They should always encourage the students to develop their ultimate performance capabilities.
- They should facilitate and encourage group work and team work to consolidate newer approach.
- They should follow Blooms taxonomy in every part of the assessment.

Bloom's Taxonomy

Level	Teacher should Check	Student should be able to	Possible Mode of Assessment
Create	Students ability to create	Design or Create	Mini project
Evaluate	Students ability to justify	Argue or Defend	Assignment
Analyse	Students ability to distinguish	Differentiate or Distinguish	Project/Lab Methodology
Apply	Students ability to use information	Operate or Demonstrate	Technical Presentation/ Demonstration
Understand	Students ability to explain the ideas	Explain or Classify	Presentation/Seminar
Remember	Students ability to recall (or remember)	Define or Recall	Quiz

GUIDELINES FOR STUDENTS

Students should take equal responsibility for implementing the OBE. Some of the responsibilities (not limited to) for the students in OBE system are as follows:

- Students should be well aware of each UO before the start of a unit in each and every course.
- Students should be well aware of each CO before the start of the course.
- Students should be well aware of each PO before the start of the programme.
- Students should think critically and reasonably with proper reflection and action.
- Learning of the students should be connected and integrated with practical and real life consequences.
- Students should be well aware of their competency at every level of OBE.

LIST OF FIGURES

UNIT-3 ANALOG MULTIPLIER AND PLL

UNIT-4 ANALOG TO DIGITAL AND DIGITAL TO ANALOG CONVERTERS

CONTENTS

1 IC Fabrication and Circuit Configuration for Linear IC

UNIT SPECIFICS

Through this unit we have discussed the following aspects:

- *Different types of Integrated Circuits;*
- *Different steps in the manufacturing of the Integrated Circuits;*
- *Different types of monolithic resistors;*
- *Discussion on monolithic capacitors and inductors;*
- *DC and AC Characteristics of an Operational Amplifier*

RATIONALE

This fundamental unit on IC Fabrication helps students to get a primary idea about the different techniques to be followed for the manufacturing of the Integrated Circuits. It explains different types of classifying Integarted Circuits. All types of fabrication steps are explained thoroughly. IC Technology important topic of electronic science that essentially deals with different processes involved in the manufacturing of the ICs.

An operational amplifier, more generally referred to as an op-amp, is a circuit that consists of three stages: the input stage, the gain stage, and the output stage. This operational amplifier can be purchased as an integrated circuit. The input stage is a differential amplifier, the gain stage offers extra voltage gain along with the necessary dc level shifting, and the output stage provides current gain along with a low output impedance. It is common practise to incorporate a feedback capacitor into the second stage in order to provide frequency adjustment. The operational amplifier (also known as an op-amp) is a device that has several terminals and is internally compensated. It was first conceived of as an integrated circuit in the year 1966. The sole signal that the operational amplifier amplifies is the difference between two input signals, and it only produces a single output signal.

PRE-REQUISITES

Electronic devices and circuits, Circuit theory

UNIT OUTCOMES

List of outcomes of this unit is as follows:

U1-O1: Describe Integrated Circuit

U1-O2: *Explain the different processing techniues*

U1-O3: *Explain Manufacturing of monolithic diodes*

U1-O4: *Apply manufacturing techniques to capacitors*

UI-05: *DC and AC Characteristics of an Operational Amplifiers*

Unit-1 Outcomes	EXPECTED MAPPING WITH COURSE OUTCOMES *(1-Weak Correlation; 2- Medium Correlation; 3-Strong Correlation)*					
	CO-1	CO-2	CO-3	CO-4	CO-5	CO-6
U1-01	3	2	-	-	1	2
U1-02	2	1	1	1	2	3
U1-03	3	1	1	1	2	1
U1-04	2	2	1	1	2	1
U1-05	1	2	2	2	3	1

1.1 Integrated Circuit

Integrated Circuit is a miniature, low cost electronic circuit consisting of active and passive components that are irreparably joined together on a single crystal chip of silicon. These are classified as,

Based on mode of operation

a. Digital IC's

b. Linear IC's

Based on fabrication

a. Monolithic IC's

b. Hybrid IC's

Based on number of components integrated on IC's

a. SSI <10 components

b. MSI <100 components

c. LSI >100 components

d. VLSI >1000 components

Digital integrated circuits (ICs) are fully functional logic networks that are comparable to simple transistor logic circuits. Examples include Gates, Counters, Multiplexers, shift registers etc. Linear integrated circuits (ICs) frequently need extra external components to function satisfactorily.

All components (active and passive) in monolithic integrated circuits (ICs) are simultaneously created by a diffusion process. The desired circuit is then created by joining these parts together via a metallization technique. Passive elements (such resistors and capacitors) and the connections between them are constructed on an insulating substrate in hybrid integrated circuits (ICs). For the integrated components, the substrate serves as a chassis. A complete circuit is then put together by connecting active parts like transistors, diodes, and monolithic integrated circuits.

1.2 Adavnatages of ICs

The following are the advanctages of ICs over discrete components

- Equipment density has risen due to miniaturisation.
- Cost savings as a result of batch processing.
- Less soldered joints, which increases system reliability.

- Improved efficiency in use.
- Matched Devices.
- Accelerated operating rates.
- Power usage reduction

1.3 Construction of a Monolithic Bipolar Transistor

The construction of a monolithic transistor includes the following steps.

1. Epitaxial growth
2. Oxidation
3. Photolithography
4. Isolation diffusion
5. Base diffusion
6. Emitter diffusion
7. Contact mask
8. Aluminium metallization
9. Passivation

1.3.1 Epitaxial Growth

The initial step in transistor fabrication is creation of the collector region which need low resistivity path. In this configuration, the N-type epitaxial layer and P-type substrate are separated by a strongly doped "N" region. This buried N+ layer offers a low resistance path to the collector contact C in the active collector region. A P-type silicon substrate with a resistivity of usually 1 Ω-cm, translating to an acceptor ion concentration of $1.4 * 10^{15}$ atoms/cm^3, is the starting point for the fabrication of an NPN transistor. The required pattern for buried layer diffusion is produced on an oxide mask. The oxide in the buried layer mask is then masked and etched after that. Now that it has diffused into the substrate, the N-type buried layer. Use of a slow-diffusing substance, such as antimony or arsenic, ensures that the buried layer will remain in place during successive diffusions. The sheet resistivity is typically approximately 20 Ω/sq, and the junction depth is typically a few microns. The P-type substrate is then formed on an epitaxial layer of weakly doped N-silicon by heating the wafer to 1200^0 C and adding a gas containing phosphorus (donor impurity). Figure illustrates the resulting structure.

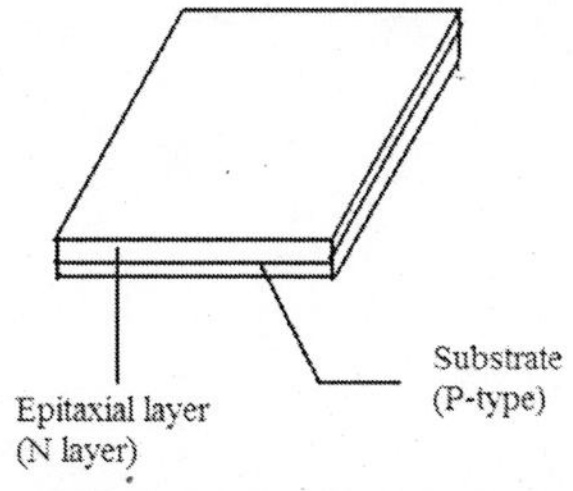

Fig.1.1 Epitaxial Growth structure

In this epitaxial layer, the diffusions that follow are carried out. The thin N-layer epitaxial layer grown over the P-type substrate serves as the base for the formation of all active and passive components. The most difficult task in processing bipolar devices is probably getting an epitaxial layer with the right thickness and doping with high crystal quality.

1.3.2 Oxidation

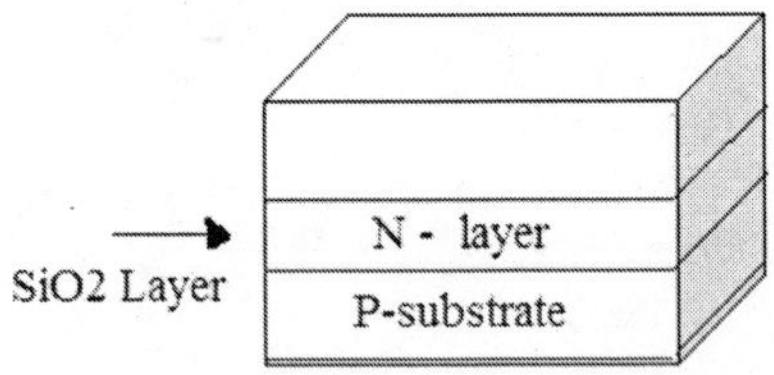

Fig.1.2. Oxidation Process

When the silicon wafer is exposed to an oxygen atmosphere at around 1,000 degrees Celsius, a thin layer of silicon dioxide (SiO2) is developed over the N-type layer, as indicated in the figure.

1.3.3. Photo Lithography

In making ICs, photolithography is mostly used to etch or remove parts of the SiO2 layer. As shown in the picture, a thin, even layer of photosensitive emulsion is first put on the oxide's surface (Photo resist). The mask is put on top of the structure. It is a black-and-white copy of the required pattern. When the photo resist under the clear part of the mask is exposed to ultraviolet light, it becomes polymerized. The mask is then taken off, and the wafer is treated chemically to get rid of the parts of the photoresist film that were not exposed. The polymerized area is allowed to harden so that it won't rust. The chip is then put in a solution of hydrofluoric acid that eats away at the oxide layer that the polymerized photoresist didn't protect. This makes holes in the SiO2 layer that allow P-type or N-type impurities to move through using the process shown in figure. After the impurities have moved out, the polymerized photoresist is taken off with sulphuric acid and abrasion.

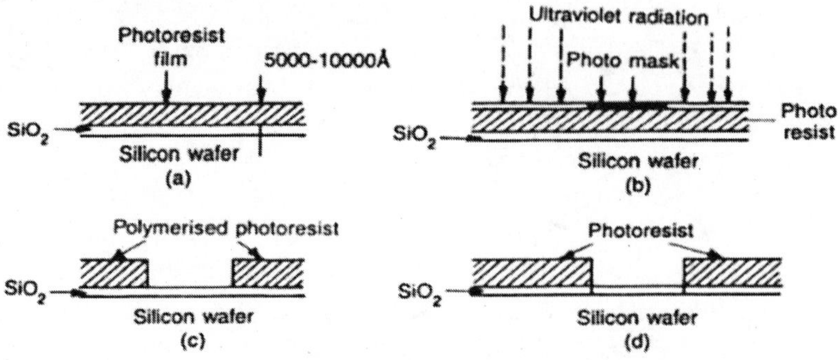

Fig.1.3.Photo Lithography Process

1.3.4 Isolation Diffusion

There are many devices on the integrated circuit. Since many devices will be made on the same IC chip, there needs to be good separation or isolation between the different parts and how they connect to each other. The most widely used techniques for isolation are,

1. PN junction Isolation
2. Dielectric Isolation

The P+ type impurities are selectively diffused into the N-type epitaxial layer so that it touches the P-type substrate at the bottom in the PN junction isolation technique. This method generated P-type moats surrounded by N-type isolation regions. When the P-substrate is held at its most negative potential, the diodes become reverse-biased, resulting in isolation between these islands. Individual components are created within these islands. This method is the most commonly used isolation method for general purpose integrated circuits because it is very cost effective.

In the dielectric isolation method, each part is surrounded by a layer of a solid dielectric, like silicon dioxide or ruby, which acts as an insulator. There is both physical and electric isolation. This method is very expensive because it takes a lot of extra steps to process, and it is mostly used to make ICs for special uses in the military and aerospace.

1.3.5. Base Diffusion

In making a bipolar transistor, one of the most important steps is making the base. The base must be in the right place so that it doesn't touch either the isolation region or the buried layer when diffusion happens. Often, the base diffusion step is also used to make diffused resistors for the circuit at the same time. The value of these resistors is determined by the conditions of diffusion and the size of the hole made by etching. The base width has a big effect on the parameters of the transistor. So, the depth and resistivity of the base junction must be tightly controlled. So that the base doesn't send electrons into the emitter, the resistivity of the base sheet should be between 200 and 500 Ohm per square. Boron is used to spread out the base of an NPN transistor in a furnace. The diffusion process is done in two steps: first, the dopants are put down at 900^0 C, and then, at about 1200^0 C, they are driven in. The drive-in is done in an environment that causes oxidation, so that oxide grows over the base region for the next steps in making the chip. Figure shows that the P-type base region of the transistor moved into the N-type island (collector region) using photolithography and isolation diffusion processes.

1.3.6 Emitter Diffusion

The final step in the fabrication of a transistor is emitter diffusion. The emitter opening must be completely contained within the base. Emitter masking not only opens windows for the emitter, but also for the contact point, allowing the emitter terminal to have a low resistivity ohmic contact path.

1.3.7 Contact Mask

After the emitter is made, holes are cut into the N-type parts where the collector and emitter terminals will be connected. Heavy concentrations of phosphorus N+ dopant are moved into these areas at the same time.

1.3.8 Metallization

Now that the active and passive devices are on the IC chip, the metal leads that will connect to the terminals of the devices need to be made. Aluminium is deposited over the entire wafer by vacuum deposition. Metal with a single layer is 1μm thick. As shown in the figure, metallization is done by evaporating aluminium over the whole surface and then selectively etching away aluminium to leave the desired interconnection and bonding pads. Metallization is done to make connections between the different parts of an IC and to put bonding pads around the edge of the IC chip so that wires can be connected later.

1.3.9 Passivation/ Assembly and Packaging

After metallization comes passivation, which is the process of putting an insulating and protective layer over the whole device. This keeps it from getting damaged by mechanical or chemical damage during the next steps of processing. Most layers are passivated with doped or undoped silicon oxide or silicon nitride, or a mixture of the two. Chemical vapour deposition (CVD) is used to put down the layer at a low enough temperature not to hurt the metallization.

1.4 Monolithic Diodes

The diode is formed in three ways . The details is as follows,

1. The Base-Emitter connection, with collector short circuited
2. The Base-Emitter Connection, with collector open circuited
3. The Collector-Base connection, with emitter open circuited

The structure of a diode depends on how well it works and what it will be used for.

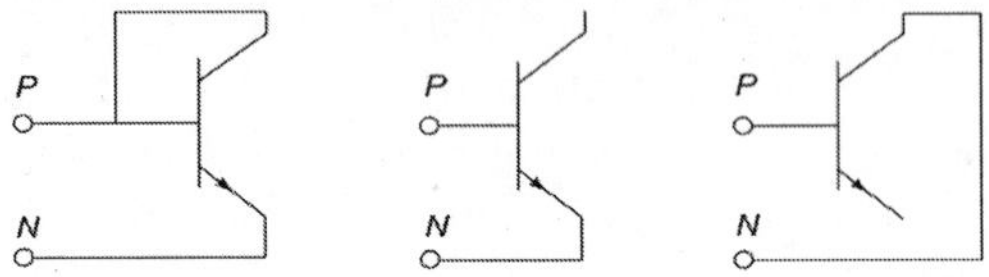

Fig.1.4.Diode Configuartions

1.5 Integrated Resistors

The commonly used methods for the fabrication of integrated resistors are

1. Diffused
2. Epitaxial
3. Pinched
4. Thin film

1.5.1 Diffused resistor

Figure shows how resistor is manufactured using diffusion process . People find this process to be very useful, so base-diffused resistors have become very popular.

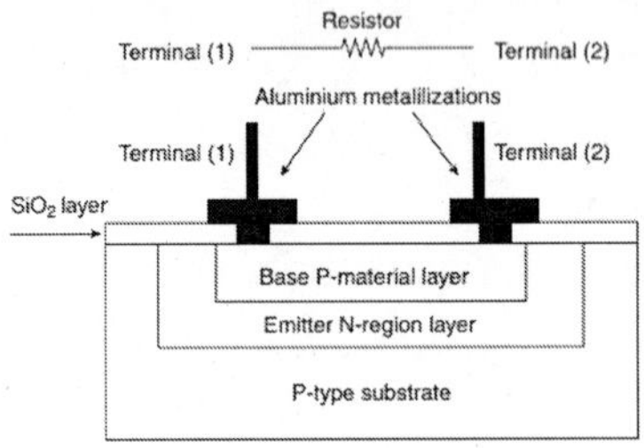

Fig.1.5.Diffused Resistor

These resistors have their limitations and can be used for the manufacturing upto some kilo ohms.

1.5.2 Epitaxial resistor

This technique is used to make high value resistances

1.5.3 Pinched Resistor

By making the diffusion regions' effective cross-sectional area smaller, the sheet resistance they offer can be raised. Most of the time, this kind of resistance is found in the base.

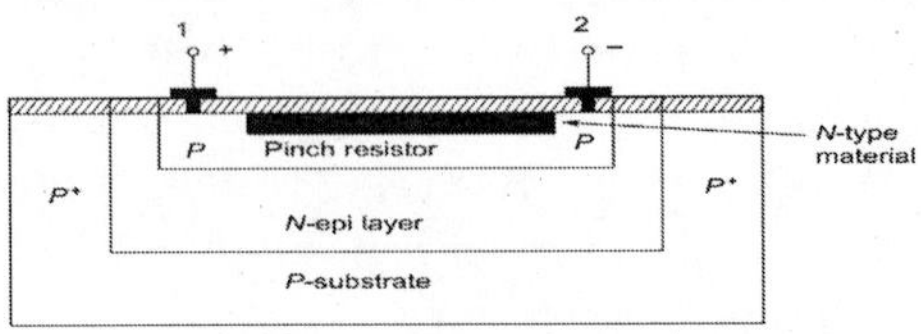

Fig.1.6.Pinched Resistor

Figure shows a resistor with a pinched base and a diffused top. It can provide resistances in the mega-ohm range in a relatively small area. In the structure shown, the diode at contact 2 is biassed in the wrong direction, so no current can flow through the N-type material. In the N-region, only a very small amount of reverse saturation current can flow. So, by making this N-shaped area in the base diffusion, the current path has been shortened or "pinched." So, as the width gets smaller, the resistance between contact 1 and contact 2 goes up. This makes it work like a pinched resistor.

1.5.4 Thin Film Resistor

Monolithic resistors can also be made using the thin film deposition technique. Vapour deposition methods are used to produce a very thin metallic film over the silicon dioxide layer, with a thickness of less than 1μm. Nichrome (NiCr) is typically employed in this technique. To produce the desired geometry and the right value of resistors, masked etching procedures are used. Aluminum metallization, as mentioned in the previous sections, is used to create ohmic connections. These resistors can be suitable to be used at high freqiencies. These resistors are more stable.

1.6 Monolithic Capacitors

In a reverse biased PN junction

$$C_T = \frac{\epsilon A}{W}$$

The capacitance is directly proportional to area(A) and inversely proportional to width of the depletion layer(W). Together with the two neighbouring conducting layers (made of P and N materials), the depletion region (which serves as a dielectric) creates a capacitor.

There are two methods of fabrication of IC capacitors.

1. PN junction with depletion region as dielectric
2. MOS capacitors

Since their performance and value range are constrained, monolithic capacitors are rarely employed in integrated circuits.

1.7 Monolithic Inductors

There are no reliable integrated inductors available. As the value of the inductance increase, cost and space also increase. Therefore, when using integrated circuits, inductors are typically avoided.

1.8 Operational amplifier:

The operational amplifier is a direct-coupled high gain amplifier that can be used from 0 to over 1MHz. Feedback is supplied to the operational amplifier in order to regulate its overall response characteristic, which includes its gain and bandwidth. The operational amplifier displays gain all the way down to zero frequency.

Figure 1.3 depicts the internal block diagram of an operational amplifier (op-amp). The dual input balanced output differential amplifier is found at the input stage of the circuit. In most cases, this stage is responsible for the majority of the voltage gain that the amplifier achieves. Additionally, it is responsible for determining the input resistance of the operational amplifier. The intermediate stage is often comprised of a second differential amplifier, which is powered by the output of the stage that came before it. The intermediate stage of the vast majority of amplifiers features dual inputs and unbalanced outputs. Because of direct coupling, the dc voltage that is produced at the output of the intermediate stage is significantly higher than the potential that is found at ground level. As a result, the level translator (shifting) circuit is utilised after the intermediate stage in order to bring the voltage down to zero volts in relation to the ground. In most cases, the output stage of the amplifier will be a push-pull complementary symmetry amplifier. The output stage amplifies the voltage swing and boosts the operational amplifier's ability to supply ground. A low output resistance is also provided by an output stage that has been appropriately constructed.

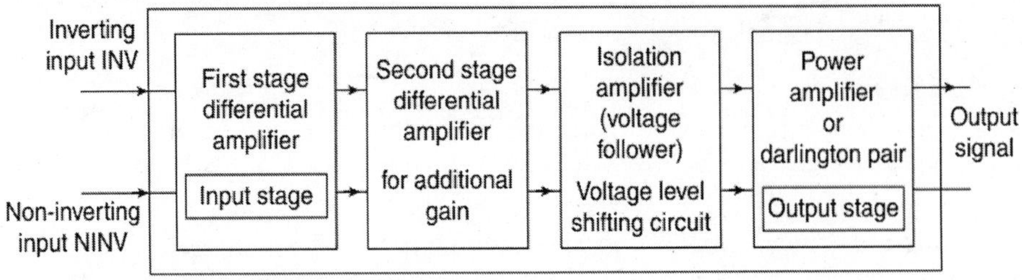

Fig.1.7.Internal diagram of an operational Amplifier

1.9 Differential Amplifier

The differential amplifier is a fundamental component that makes up an operational amplifier (op-amp). A differential amplifier's job is to amplify the difference between the signals that are sent into it from two separate sources. Both the Q1 and Q2 transistors share the same qualities in their construction. Both of the circuits have the same resistance, therefore $R_{E1} = R_{E2}$ and $R_{C1} = R_{C2}$, and the magnitude of the $+V_{CC}$ voltage is the same as the magnitude of the $-V_{EE}$ voltage. When measuring these voltages, ground is always considered to be the reference point.

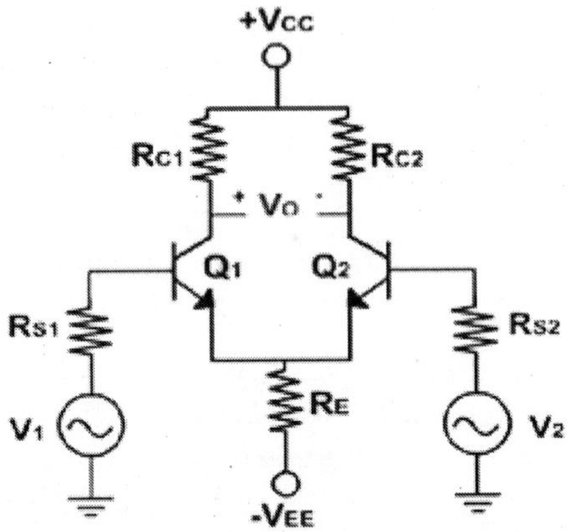

Fig.1.8. Differential Amplifier

The two circuits need to be connected in the manner depicted in figure 1.4 in order to create a differential amplifier. Due to the fact that they are identical, the two supply terminals marked +VCC and -VEE have been combined into one. In addition to this, the two emitters are linked together, and the parallel combination of RE1 and RE2 is switched out for a resistance RE. Both the base of Q1 and the base of Q2 are where the input signals v1 and v2 are applied respectively. Between the two collectors is where the output voltage is measured. Because the resistances of the collectors are identical, the equation RC = RC1 + RC2 is used to represent them.

When both of the inputs have the same value, the voltage at the output should, ideally, be zero. The output voltage with the polarity displayed appears when the value of v1 is higher than the value of v2. When v1 is lower than v2, the polarity of the output voltage will flip from positive to negative. Different configurations have been utilised for each differential amplifier.

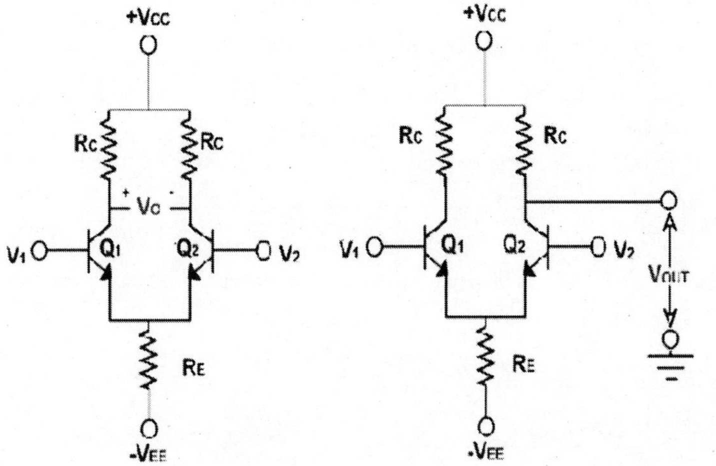

Fig1.9. Dual input, balanced output differential amplifier. And Dual input, unbalanced output differential amplifier.

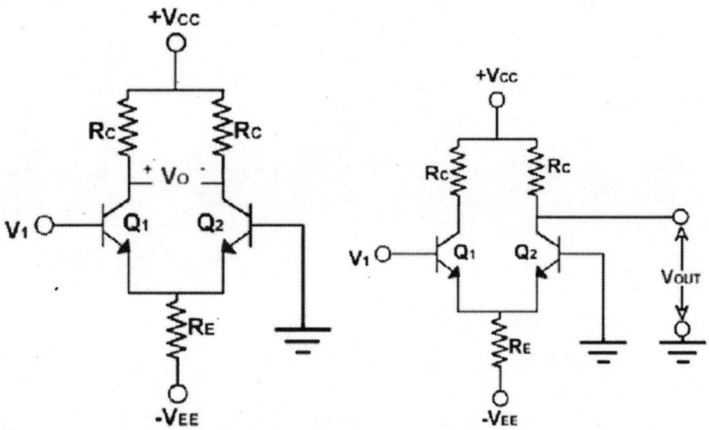

Fig.1.10 Single input, balanced output differential amplifier and Single input, unbalanced output differential amplifier.

The four differential amplifier configurations are following:

1. Dual input, balanced output differential amplifier.
2. Dual input, unbalanced output differential amplifier.
3. Single input balanced output differential amplifier.
4. Single input unbalanced output differential amplifier.

In fig, the number of input signals and output voltage measurement characterise these combinations. Dual input means two input signals, single input means one. When measured between two collectors, the output voltage is balanced since both collectors are at the same dc potential w.r.t. ground.

Unbalanced output is measured at one collector w.r.t. ground.

1.10 Dual Input, Balanced Output Differential Amplifier

The diagram of the circuit can be found here. The two inputs, V_1 and V_2, are connected to the bases of the transistors Q_1 and Q_2, respectively. The output voltage is determined by measuring the potential difference between the two collectors, C_1 and C_2, which both have the same dc potential.

1.10.1 D.C. Analysis:

In order to acquire the operating point for the differential amplifier (ICQ and VCEQ), the dc equivalent circuit must first be formed. This is done by bringing the input voltages V1 and V2 down to zero, as shown in figure.

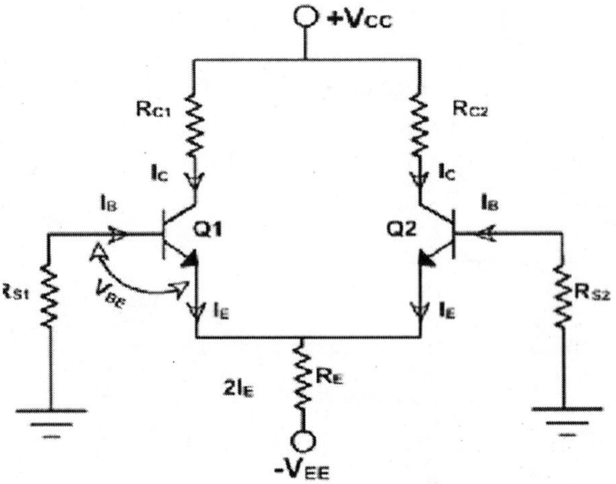

Fig 1.11 Differential Amplifier for DC Analysis

Due to the fact that RS1 = RS2, the symbol RS is used to signify the internal resistances of the input signals. Because both of the separate amplifier's emitter biased parts are symmetrical in all aspects, the operating point for only one of the sections needs to be established. The second transistor, Q2, can have its ICQ and VCEQ set to the same values as the first transistor, Q1. KVL being applied to the base emitter loop of transistor Q1 in the circuit and is given by,

$$R_S I_B + V_{BE} + 2I_E R_E = V_{EE}$$

We have, $I_B = \frac{I_E}{\beta_{dc}}$ and $I_C \approx I_E$ then the above equation reduces to

$$R_S \frac{I_E}{\beta_{dc}} + V_{BE} + 2I_E R_E = V_{EE}$$

$$I_E = \frac{V_{EE}-V_{BE}}{\frac{R_S}{\beta_{dc}}+2R_E} \approx \frac{V_{EE}-V_{BE}}{2R_E} \text{ for } \beta_{dc} \gg 1$$

From the collector side,

$V_C = V_{CC} - I_C R_C$ and $V_{CE} = V_C - V_E$

$= V_{CC} - I_C R_C + V_{BE}$

From the two equations V_{CEQ} and I_{CQ} can be determined. This dc analysis is applicable for all types of differential amplifier.

1.10.2 A.C. Analysis :

Replacing BJT with r_e model the differential amplifier reduces to,

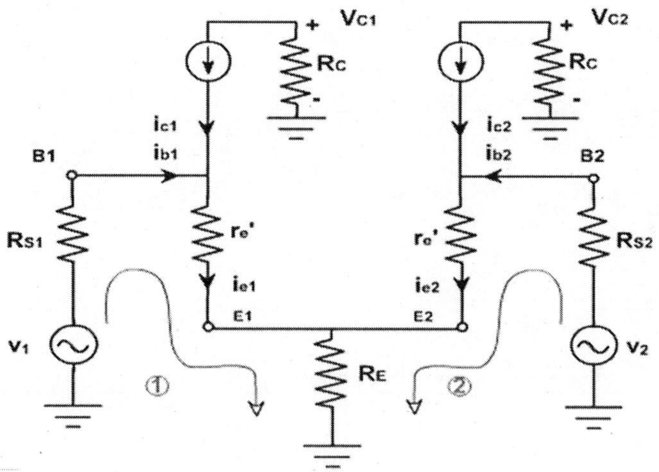

Fig 1.12.Differential Amplifier A.C Analysis

Given that the two dc emitter currents are equivalent to one another. Therefore, the resistances r_{e1} and r_{e2} are the same and are collectively referred to as r_e. This voltage across each collector resistance is shown to be 180 degrees out of phase in relation to the voltages v_1 and v_2 that are being input. There is no difference between this and the CE configuration. Figure illustrates the polarity of the output voltage for your reference. Even if both collectors are believed to have a negative relationship with ground, it is presumed that the collector C_2 has a greater favourable relationship with collector C_2. The output voltage is given by,

$$V_0 = V_{C2} - V_{C1} = R_c(i_{e1} - i_{e2})$$

$$i_{e1} = \frac{(r_e + R_E)V_1 - R_E V_2}{(r_e + R_E)^2 - {R_E}^2} = \frac{(r_e + R_E)V_1 - R_E V_2}{r_e(r_e + 2R_E)}$$

$$i_{e2} = \frac{(r_e + R_E)V_2 - R_E V_1}{(r_e + R_E)^2 - {R_E}^2} = \frac{(r_e + R_E)V_2 - R_E V_1}{r_e(r_e + 2R_E)}$$

Substituting and simplifying the expression for the output voltage will be,

$$V_0 = \frac{R_c}{r_e}(V_1 - V_2)$$

Differential Input Resistance:

Differential input resistance is defined as the equivalent resistance that would be measured at either input terminal with the other terminal grounded. This resistance can be thought of as the "normal" resistance that would be measured at either input terminal. This indicates that the value of the input resistance Ri1 observed from the input signal source V_1 should be computed with the signal source V_2 set to zero.

$$R_{i1} = \left.\frac{v_1}{i_{b1}}\right|_{v2=0} = \left.\frac{\beta v_1}{i_{e1}}\right|_{v2=0} \approx 2\beta r_e$$

In a similar manner, the input signal V_1 was made to have a value of zero in order to estimate the input resistance Ri2 observed from the source of the input signal V_2. Due to their extremely low values, the R_{S1} and R_{S2} resistances are neglected.

$$R_{i2} = \left.\frac{v_2}{i_{b2}}\right|_{v1=0} = \left.\frac{\beta v_2}{i_{e2}}\right|_{v1=0} \approx 2\beta r_e$$

Output Resistance:

Output resistance is the equivalent ground-to-output terminal resistance. The output resistance R_{O1} between collector C_1 and ground equals the collector resistance R_C. The output resistance R_{O2} measured at C_2 relative to ground matches the collector resistor R_C.

$R_{O1} = R_{O2} = R_C$

1.11 DC Characteristics of Operational Amplifier

Input Offset Voltage:

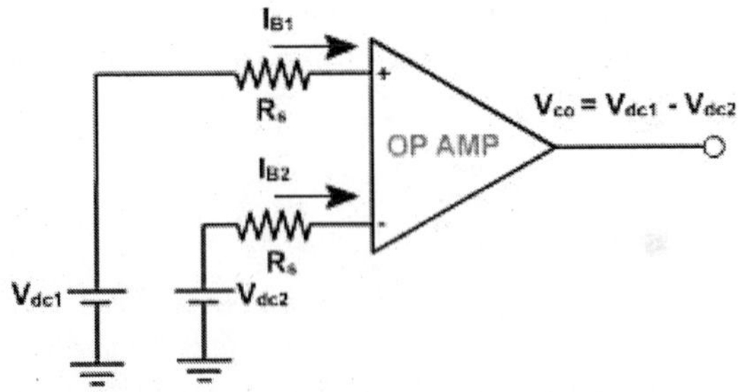

Fig 1.13 Input offset voltage

If nothing is connected to the inverting and non-inverting ends of the op-amp from the outside, the output must be zero. So, if $V_i = 0$, $V_o = 0$. But because $+V_{cc}$ and $-V_{cc}$ are used as biasing supply voltages, the op-amps draw a finite bias current, and because the differential amplifier design is not symmetric, the output will not be zero. This is called "offset." Since V_o must be zero when V_i is zero, an input voltage must be used to cancel the output error and make Vo equal to zero. It's called "input offset voltage." Input offset voltage (V_{io}) is the voltage that needs to be put between the two input lines of an OPAMP for the output voltage to be zero. Figure 1.22 shows that the output is zero when two dc voltages are put into the input lines. $V_{io} = V_{dc1} - V_{dc2}$.

The dc voltages V_{dc1} and V_{dc2} are denoted by Rs, which stands for the source resistance. The difference between Vdc1 and Vdc2 is denoted by V_{io}. It could have a positive or a negative impact. The value of V_{io} can reach a maximum of 6 mV when using a 741C OPAMP. It indicates that a voltage of less than six millivolts must be applied to one of the inputs in order to get the output offset voltage down to zero. Because of the closer match between the transistors in a differential amplifier, its performance improves in direct proportion to the input offset voltage's magnitude.

Input offset Current:

In theory, the input impedance of an ideal operational amplifier (op-amp) is infinite. However, in practise, this is not the case. Therefore, the IC receives its current from the source, regardless of how low that current may be. The term for this phenomenon is input offset current Iio. The difference between the currents flowing into the inverting and non-inverting terminals of a balanced amplifier is the input offset current, or Iio. This difference is illustrated in figure 1.22.$I_{io} = | I_{B1} - I_{B2} |$.The highest I_{io} for the 741C is 200 nA. As the fit between two input ports gets better, the difference between I_{B1} and I_{B2} gets smaller, which means the value of I_{io} goes down even more. I_{io} is 6 nA for a precision OPAMP 741C.

Input Bias Current:

The average amount of current flowing through a balanced amplifier's input terminals

is the value that is referred to as the input bias current, or IB. i.e.

IB = (IB1 + IB2) / 2

For ideal op-amp IB=0. For 741C IB(max) = 700 nA and for precision 741C IB = ± 7 nA

1. Differential Input Resistance: (Ri)

R_i is the equivalent resistance that can be measured at either the inverting or the non-inverting input terminal with the other terminal being grounded. This resistance can be measured in either direction. The input resistance of the 741C is measured to be 2 M, which is considered to be on the higher end. It could be as high as 1000 G ohm for certain OPAMPs.

Input Capacitance: (Ci)

C_i is the equivalent capacitance that can be measured at either the inverting or non-inverting terminal when the other terminal is connected to ground. This capacitance can be measured at either terminal. For the 741C, a value of C_i of 1.4 pf is considered to be average.

Offset Voltage Adjustment Range:

741 OPAMPs have the capability to null out offset voltage. For this reason, pins 1 and 5 have been given the offset null designation. It is possible to achieve this by connecting a pot with 10 K ohm resistance between 1 and 5.

The output offset voltage can be brought down to zero volts (assuming that the inputs are grounded) by adjusting the potentiometer. The range within which the input offset voltage can be modified by adjusting the value of the 10 K pot is referred to as the offset voltage adjustment range. The offset voltage adjustment range for the 741C is between -15 and 15 millivolts.

Input Voltage Range :

The input voltage range of a differential amplifier refers to the range of a common mode input signal for which the amplifier continues to remain linear. It is put to use to ascertain how well the inverting and non-inverting input terminals match up with one another. In the case of the 741C, the maximum range for the input common mode voltage is a minimum of 13 volts. This indicates that the common mode voltage that is delivered to both input terminals can range anywhere from -13V all the way up to +13V at its highest point.

Common Mode Rejection Ratio (CMRR).

CMRR is defined as the ratio of the differential voltage gain Ad to the common mode

voltage gain ACM

$CMRR = A_d / A_{CM}$.

The typical CMRR for a 741C is 90 dB. The CMRR measures how well two input terminals match one another, and a greater number indicates better matching and lower common mode voltage at the output.

2. Supply voltage Rejection Ratio: (SVRR)

The input offset voltage SVRR is the ratio of the input offset voltage change to the supply voltage change.

$SVRR = \Delta V_{io} / \Delta V$

Where ΔV is the change in the input supply voltage and ΔV_{io} is the corresponding change inthe offset voltage.For the 741C, SVRR = 150 μ V / V.For 741C, SVRR is measured for both supply magnitudes increasing or decreasing simultaneously, with R_3= 10K. For same OPAMPS, SVRR is separately specified as positive SVRR and negative SVRR.

3. Large Signal Voltage Gain:

Since the OPAMP amplifies difference voltage between two input terminals, the

voltage gain of the amplifier is defined as,

$$Volatge\ Gain = \frac{Output\ Voltage}{Differential\ Input\ Voltage} = \frac{V_O}{V_{id}}$$

Large signal voltage gain is so named because the amplitude of the amplified signal is substantially larger than the original input signal. The normal voltage gain for a 741C is 200,000.

Output voltage Swing:

The power compliance of the AC PP is the peak-to-peak voltage that an OPAMP can make without being clipped. Since the output at rest should be zero, the ac output voltage can go either up or down. This also shows the numbers of the OP-AMP's positive saturation voltage and its negative saturation voltage. For a given source voltage, +VCC and -VEE, the output voltage never goes above these limits. It is about 13 V for a 741C.

Output Resistance: (RO)

R_O is the resistance that can be measured between the OPAMP's output port and ground. The 741C OPAMP has a 75-ohm resistance.

Output Short circuit Current :

In some situations, an OPAMP may be used to move a load with a resistance close to zero. Even though it has a 75-ohm output impedance, it can't give large currents. Since an OPAMP is a low-power gadget, it can only send out a limited amount of current. The 741C can only give a short circuit current of 25mA at most.

Supply Current:

IS is the amount of current that the OP-AMP takes from the source. The current going into the 741C OPAMP is 2.8 m A.

Power Consumption:

Power consumption, often known as PC, refers to the amount of power that an OPAMP must draw while in a quiescent state (when V_{in} is equal to 0 V) in order for it to function as intended. The 741C uses up a total of 85 milliwatts of power throughout its operation.

1.12 AC Charactersitics of an OpAmp

Gain Bandwidth Product:

When the open loop voltage gain is lowered to 1, the gain bandwidth product of the OPAMP is equal to the bandwidth of the OPAMP. Based on the graph of open loop gain against frequency at the frequency of 1 MHz, as illustrated in figure 1.24, the gain of the 741C OPAMP drops down to 1. The cutoff frequency for the mid band voltage is 10 Hz, while the gain for the mid band is 100,000.

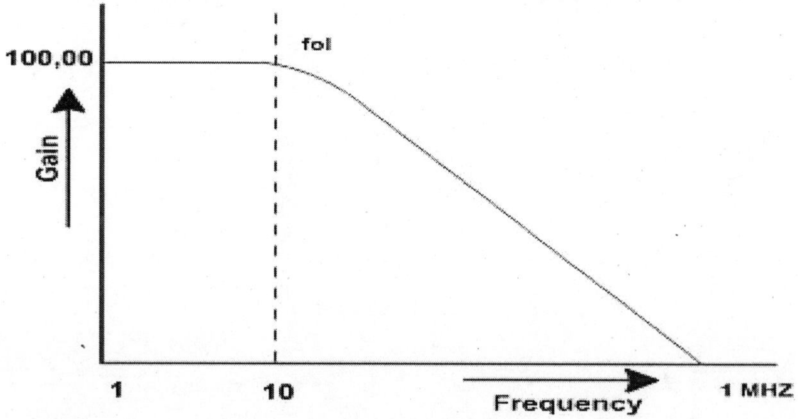

Fig.1.14 Band width of OP-AMP

Slew Rate: The highest rate of change of output voltage per unit of time under big signal conditions is referred to as the slew rate, and it is denoted as volts per microsecond (sec). Consider the flow of current through a capacitor while trying to grasp this concept.

$$\frac{i}{C} = \frac{dV}{dt}$$

If 'i' is bigger, the capacitor charges faster. If 'i' can only go up to I_{max}, then the rate of change can also only go up to Imax. The slew rate shows how quickly the output of an OP-AMP can change when the input frequency changes but the input amplitude stays the same. The slew rate changes as the voltage gain does, and it is usually given at a gain of 1. If the slope minimum is higher than the slew rate, distortion happens. The 741C has a low slew rate of 0.5V/μS, which makes it hard to use in situations with a high frequency.

Input Offset Voltage and Current Drift:

It is also called the average temperature coefficient of input offset voltage or input offset current. The ratio of the change in input offset voltage to the change in temperature is the input offset voltage drift, which is written as ΔV /° C. Input offset voltage drift = (ΔVio /ΔT). Similarly, input offset current drift is the ratio of the change in input offset current to the change in temperature. Input offset current drift = (ΔIio / ΔT).For 741C,ΔVio / ΔT = 0.5 V / C. Δ Iio/ ΔT = 12 pA / C

1.13 PIN DIAGRAM OF 741-OP AMP

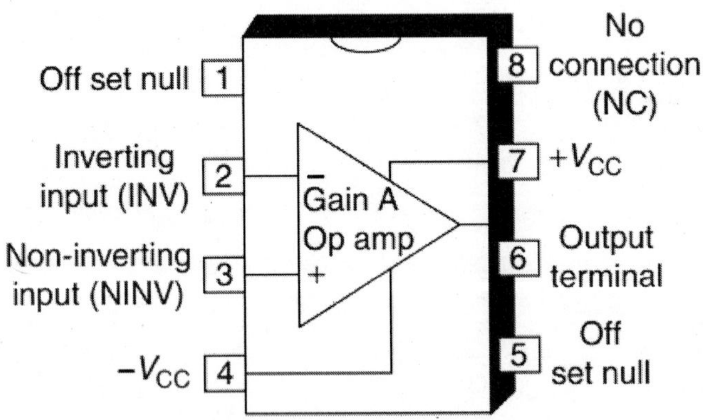

FEATURES OF 741 OP-AMP:

1. No External frequency compensation is required
2. Short circuit Protection
3. Off Set Null Capability
4. Large Common mode and differential Voltage ranges
5. Low Power Dissipation
6. No-Latch up Problem
7. 741 is available in three packages:- 8-pin metal can, 10-pin flat pack and 8 or 14-pin DIP.

1.14 MODES OF OPERATION OF OP-AMP

There following are the 2 modes in which an op-amp operates:

1. open loop mode
2. closed loop mode

1.14.1 Open loop OPAMP mode:

When discussing amplifiers, the phrase "open loop" refers to the fact that there is no connection of any kind between the input and output terminals of the device. That is to say, the signal at the output is not incorporated in any way, shape, or form into the signal at the input. When used in open loop arrangement, the OPAMP performs the duties of a high gain amplifier. There are three different open loop configurations for the OPAMP.

The Differential Amplifier:

input signals vin1 and vin2 are applied to the positive and negative input terminals of the open loop differential amplifier. This setup is known as a differential amplifier because the OPAMP enhances the difference between the two input signals. The OPAMP amplifies input signals that are both ac and dc. Normally, the input resistance Ri is much greater than the source resistances Rin1 and Rin2. Therefore, it is safe to infer that there is no voltage loss between these resistances. Therefore, $v_1 = v_{in1}$ and $v_2 = v_{in2}$. $v_o = A_d (v_{in1} - v_{in2})$ where, A_d is the open loop gain.

The Inverting Amplifier:

The term "inverting amplifier" refers to an amplifier in which the input signal is solely applied to the inverting terminal, while the non-inverting terminal is grounded.. This configuration is shown in fig.

v1= 0, v2 = vin. vo = -Ad vi

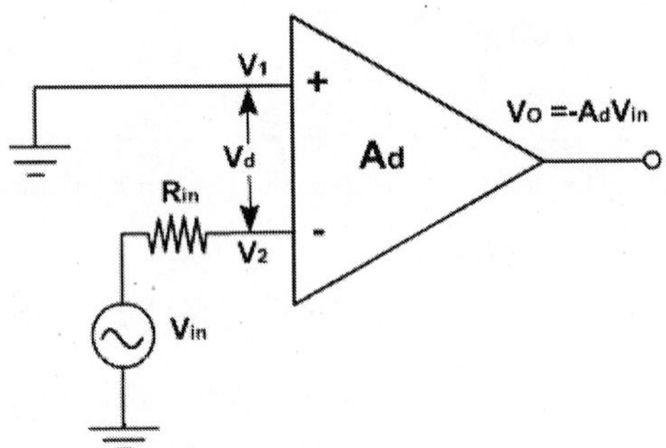

Fig 1.15 Inverting Amplifier

When you see a negative sign, it means that the voltage at the output is 180 degrees out of phase with the voltage at the input, or that it has the opposite polarity. As a result, the signal from the input is both amplified and reversed.

The non-inverting amplifier:

As can be seen in the figure, the input voltage is connected to the terminals that are not inverting, and the inverting terminal is connected to ground in this design.

$v_1 = +v_{in}$, $v_2 = 0$ $v_o = +A_d v_{in}$

This means that the input voltage is amplified by A_d and there is no phase reversal at the output.

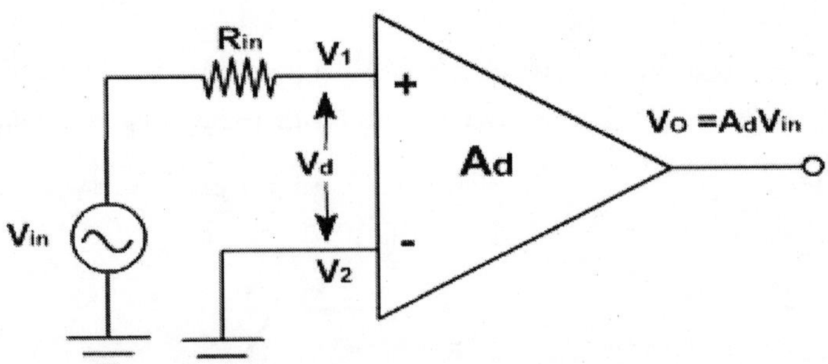

Fig 1.16 Non Inverting Amplifier

In all of the setups, if the input signal is just a little bit higher than zero, the output goes all the way up to the saturation level. This is because the gain is so big. So, when the OPAMP is working in open-loop, its output is either negative saturation or positive saturation, or it changes between the two. So, open loop op-amp is not used in linear applications.

UNIT SUMMARY

- Integrated circuits are solid-state devices and parts that have been shrunk down and built on a single crystal of silicon.
- An IC has a number of benefits, including enhanced performance, durability, and toughness, as well as low power consumption and lower cost.
- ICs can be classified into SSI, MSI, LSI, VLSI, ULSI and GSI based on device density, and analog and digital based on the types of signals they process.
- IC production uses a variety of isolation process approaches, including PN-junction and dielectric isolation techniques.
- Fabrication of capacitors of value 0.3 to 0.8 nF/mm^2 are possible in IC Technology
- Inductors can not be fabricated easily in integrated form
- Diffused, Epitaxial, pinched, thin film resistors are most commonly used one.

Multiple Choice Questions

1. The following is a valid diode connection

   a)BE junction connected, Collector open

   b)BE junction connected, Collector shorted

   c) Both a and b

   d) None of the above

[Ans.(c)]

2. The capacitance of a diode is

   a)inversely proportional to depletion layer thickness

   b)Directly proportioal to depletion layer thickness

   c)Logarithmic relation

   d)Sine relation

[Ans.(a)]

3. Function of differential amplifier
    a. Noise reduction (b) Amplification of DC Voltages
    b. Amplification of AC Voltages (d) Quick response

[Ans. (a)]

4. Operational amplifier amplifies the following signals
    a. AC signals (b) DC Signals
    b. Both AC and DC signals (d) Noise

[Ans. (c)]

5. Last stage inside the operational amplifier IC 741 is
    a. Voltage amplifier (b) Differential amplifier
    b. Current amplifier (d) Power Amplifier
    (e) Voltage follower

[Ans. (d)]

Short and Long Answer Type Questions

1. Define an Integrated Circuit? What are the advantages of an IC over discrete components?
2. Enumerate the steps involved in the manufacturing of an IC.
3. Explain the difference between epitaxial and crystal growing techniques?
4. Define and explain about isolation techniques normally used
5. With necessary diagrams, explain the steps involved in the fabrication of an IC transistor
6. Discuss about monolithic Capacitors and inductors
7. Explain the limitations of various integrated resistors

REFERENCES AND SUGGESTED READINGS

1. Coughlin, R. F. and Frederick F. Driscoll (2001). Operational Amplifiers and Linear Integrated Circuits, Sixth Edition. Upper Saddle River, NJ: Prentice Hall.
2. Dailey, Denton J. (1989) Operational Amplifiers and Linear Integrated Circuits: Theory and Applications. New York: The McGraw-Hill Companies.
3. Franco, Sergio (2002). Design with Operational Amplifiers and Analog Integrated Circuits.New York: The McGraw-Hill Companies.
4. Gayakwad, Ramakant A. (1999). Op-Amps and Linear Integrated Circuits. Fourth edition.Upper Saddle River, NJ: Prentice Hall Inc.
5. Jacob, J. Michael (1996) Applications and Design with Analog Integrated Circuits. Upper Saddle River, NJ: Prentice Hall Inc.
6. *Ramakanth A. Gayakwad* (2000),Op-Amps and Linear Integrated Circuits -, *Prentice-Hall, 4th edition.*

7. S.Salivahanan,V S Kanchana Bhaskaran(2015),Linear Integrated Circuits, Second Edition, Tata McGrawa Hill Edition.
8. B.Visweswara Rao(2015), Linear Integrated Circuits, First Edition, Pearson India Education Services Pvt. Ltd

Dynamic QR Code for Further Reading

2 Applications of Operational Amplifiers

UNIT SPECIFICS

Through this unit we have discussed the following aspects:

- *Difference between Inverting and Noninverting amplifiers;*
- *Different types of filter Circuits;*
- *Different types of signal processing circuits;*
- *Discussion on Instrumentation Amplifier;*
- *Voltage to current and current to voltage converters;*

RATIONALE

The operational amplifier, sometimes known as an op-amp, was first designed as a solution to a problem that was being faced by analogue computer designers. An operational amplifier (or op-amp for short) is a high-gain, direct-coupled amplifier. The feedback components that are connected to the circuit externally allow for regulation of the voltage gain. The operational amplifier can be put to use in amplifier and signal processing applications that span frequency ranges from dc to several MHz. Circuits for operational amplifiers can be constructed using a wide variety of active devices in various configurations. In spite of this, integrated circuit technology has been extremely effective in producing low-cost, high-performance, and adaptable op-amps in a monolithic form. As a result, the op-amp has become a widely accepted building component of modern signal processing and conditioning circuits.

Applications for negative feedback amplifiers make use of IC op-amps because of their low cost, versatility, and ease of use. IC op-amps also find uses in waveshaping, filtering, and the solution of mathematical operations. In this chapter, we will go over some of the more typical applications.

PRE-REQUISITES

Electronic devices and circuits, Circuit theory

UNIT OUTCOMES

List of outcomes of this unit is as follows:

U2-O1: Describe Sign changer and scale changer

U2-O2: Explain the different filters

U2-O3: Explain about Instrumentation Amplifiers

U2-O4: Discuss about voltage to current and current to voltage converters

U2-O5: Discuss about log and antilog amplifiers

Unit-2 Outcomes	**EXPECTED MAPPING WITH COURSE OUTCOMES** (*1-Weak Correlation; 2- Medium Correlation; 3-Strong Correlation*)					
	CO-1	**CO-2**	**CO-3**	**CO-4**	**CO-5**	**CO-6**
U2-01	1	3	1	1	2	3
U2-02	1	3	2	2	2	1
U2-03	1	3	2	2	2	1
U2-04	-	3	2	2	2	
U2-05	-	3	2	2	2	3

2.1 Inverting Amplifier Configuration

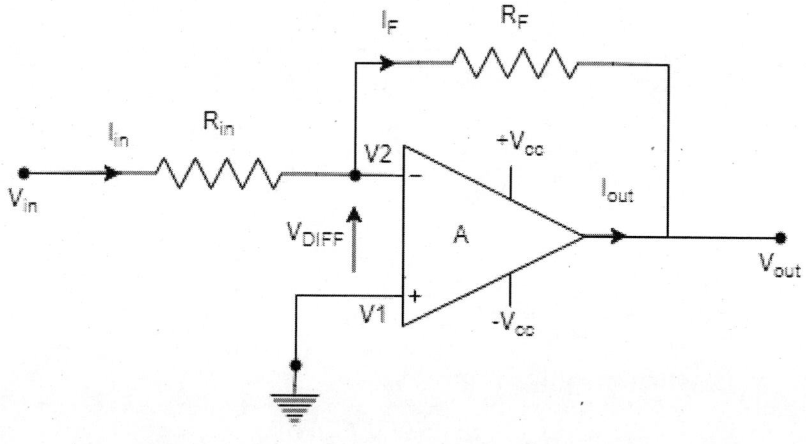

Fig.2.1 : Inverting amplifier with feedback.

In this Inverting Amplifier circuit, feedback is linked to the operational amplifier to make a closed loop. There are two important rules to know about inverting amplifiers for ideal op-amps: "no current flows into the input terminal" and "V1 equals V2" (in real-world op-amps, both of these rules are broken). This is because the point where the input and return signal (X) meet is at the same potential as the positive (+) input, which is zero volts or ground. This point is called a "virtual earth." Because of this virtual earth node, the input resistance of the amplifier is equal to the value of the input resistor, Rin. The ratio of the two external resistors can be used to set the closed loop gain of the inverted amplifier.

We said above that there are two very important rules to remember about Inverting Amplifiers or any operational amplifier for that matter and these are.

1. No Current Flows into the Input Terminals
2. The Differential Input Voltage is Zero as $V_1 = V_2 = 0$ (Virtual Earth)

Then by using these two rules we can derive the equation for calculating the closed-loop gain of an inverting amplifier, using first principles. Current (i) flows through the resistor network as shown.

$$i = \frac{V_{in} - V_{out}}{R_{in} + R_f}$$

Therefore, $$i = \frac{V_{in} - V_2}{R_{in}} = \frac{V_2 - V_{out}}{R_f}$$

$$i = \frac{V_{in}}{R_{in}} - \frac{V_2}{R_{in}} = \frac{V_2}{R_f} - \frac{V_{out}}{R_f}$$

So, $$\frac{V_{in}}{R_{in}} = V_2\left[\frac{1}{R_{in}} + \frac{1}{R_f}\right] - \frac{V_{out}}{R_f}$$

And as,

$$\mathrm{i} = \frac{\mathrm{Vin} - 0}{\mathrm{Rin}} = \frac{0 - \mathrm{Vout}}{\mathrm{Rf}} \text{ and } \left[\frac{\mathrm{Rf}}{\mathrm{Rin}}\right] = \frac{0 - \mathrm{Vout}}{\mathrm{Vin} - 0}$$

Voltage Gain of an Inverting Amplifier is given as

$$\mathrm{Av} = \frac{\mathrm{Vout}}{\mathrm{Vin}} = -\frac{\mathrm{Rf}}{\mathrm{Rin}}$$

Then, the Closed-Loop Voltage Gain of an Inverting Amplifier is given as

$$Av = \frac{Vout}{Vin} = -\frac{Rf}{Rin}$$

and this can be transposed to give Vout as: $Vout = \left[-\frac{Rf}{Rin}\right] * Vin$

The negative sign in the equation indicates an inversion of the output signal with respect to the input as it is 180° out of phase. This is due to the feedback being negative in value.

2.2 The Non-inverting Amplifier

A non-inverting amplifier is the second fundamental configuration of a circuit for an operational amplifier. This configuration does not flip the signal. Because the input voltage signal, Vin, is applied directly to the non-inverting (+) input terminal in this configuration, the output gain of the amplifier becomes "Positive" in value. This is in contrast to the "Inverting Amplifier" circuit that we examined in the previous lesson, which had an output gain that was negative in value. As a consequence of this, the signal at the output is said to be "in-phase" with the signal at the input.

Again providing negative feedback, the R_f - R_2 voltage divider network is used to achieve feedback control of the non-inverting amplifier. This is done by applying a little portion of the output voltage signal back to the inverting (-) input terminal. This closed-loop architecture results in a non-inverting amplifier circuit that has very strong stability, extremely high input impedance, R_{in} approaching infinity, because there is no current flowing into the positive input terminal, (ideal conditions), and low output impedance, Rout as illustrated below.

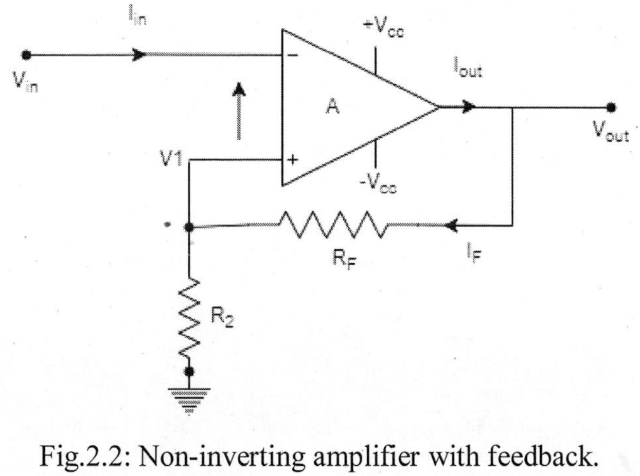

Fig.2.2: Non-inverting amplifier with feedback.

2.2.1 Non-inverting Amplifier Configuration

According to what is stated in the article about the inverting amplifier, "no current flows into the input" of the amplifier, and "V1 equals V2." This was due to the fact that the junction of the input signal and the feedback signal, shown by the voltage V1, was at the same potential. The junction can be thought of as a "virtual earth" summing point. Because of this virtual ground node, the resistors R and R2 create a simple potential divider network across the non-inverting amplifier. The ratios of R2 and R determine the voltage gain of the circuit, as illustrated in the following diagram

2.2.2 Equivalent Potential Divider Network

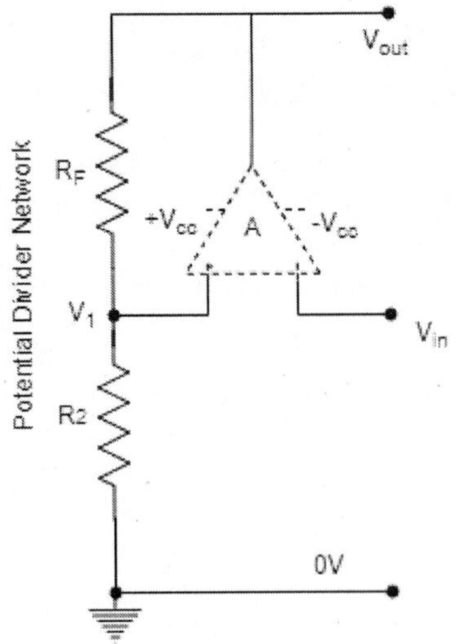

Fig2.3: potential divider in non-inverting op-amp

From the fig.2.3 using the formula to calculate the output voltage of a potential divider network, we can calculate the closed-loop voltage gain (Av) of the Non-inverting Amplifier as follows:

$$V_1 = \frac{R_2}{R_2 + R_f} * V_{out}$$

$$\text{Ideal Summing Point : } V_1 = V_{in}$$

$$\text{Voltage Gain } A_v = \frac{V_{out}}{V_{in}}$$

$$\text{Then, } A_v = \frac{V_{out}}{V_{in}} = \frac{R_2 + R_f}{R_2}$$

$$\text{Transpose to give : } A_v = \frac{V_{out}}{V_{in}} = 1 + \frac{R_f}{R_2}$$

$$\text{Then,}\quad A_v = \frac{V_{out}}{V_{in}} = \frac{R_2+R_f}{R_2}$$

We can see from the equation above, that the overall closed-loop gain of a non- inverting amplifier will always be greater but never less than one (unity), it is positive in nature and is determined by the ratio of the values of R_f and R_2. If the value of the feedback resistor R_f is zero, the gain of the amplifier will be exactly equal to one (unity). If resistor R_2 is zero the gain will approach infinity, but in practice it will be limited to the operational amplifiers open-loop differential gain, (A_o).

2.3 Voltage Follower (Unity Gain Buffer)

If we made the feedback resistor, R_f equal to zero, ($R_f = 0$), and resistor R_2 equal to infinity, ($R_2 = \infty$) as shown in fig , then the circuit would have a fixed gain of "1" as all the output voltage would be present on the inverting input terminal (negative feedback). This would then produce a special type of the non-inverting amplifier circuit called a Voltage Follower or also called a "unity gain buffer".

As the input signal is connected directly to the non-inverting input of the amplifier the output signal is not inverted resulting in the output voltage being equal to the input voltage, $V_{out} = V_{in}$. This then makes the voltage follower circuit ideal as a Unity Gain Buffer circuit because of its isolation properties as impedance or circuit isolation is more important than amplification while maintaining the signal voltage. The input impedance of the voltage follower circuit is very high, typically above 1MΩ as it is equal to that of the operational amplifiers input resistance times its gain (R_{in} x A_o). Also its output impedance is very low since an ideal op-amp condition is assumed.

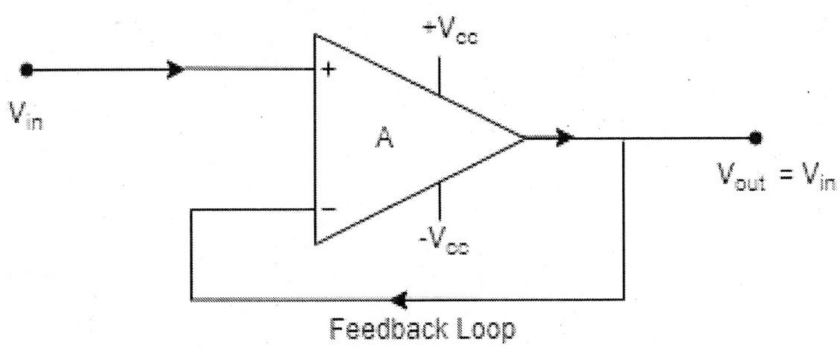

Fig.2.4 : voltage follower

In this non-inverting circuit configuration, the input impedance R_{in} has increased to infinity and the feedback impedance R_f reduced to zero. The output is connected directly back to the negative inverting input so the feedback is 100% and Vin is exactly equal to V_{out} giving it a fixed gain of 1 or unity. As the input voltage V_{in} is applied to the non-inverting input the gain of the amplifier is given as:$V_{out} = A_v + V_{in}$

$V_{in} = V + and\ V_{out} = V$-

One final thought, the output voltage gain of the voltage follower circuit with closed loop gain is Unity, the voltage gain of an ideal operational amplifier with open loop gain (no feedback) is Infinite. Then by carefully selecting the feedback components we can control the amount of gain produced by an operational amplifier anywhere from one to infinity.

2.4 Instrumentation Amplifier:

The monitoring and regulation of the environment's physical conditions play an essential role in a wide variety of commercial and domestic applications. For instance, taking readings of the temperature and humidity within a dairy or meat processing factory gives the operator the ability to make any necessary modifications to ensure that the product quality is maintained. In a similar vein, in order to manufacture a specific variety of plastic, the temperature management of the plastic furnace must be extremely accurate.

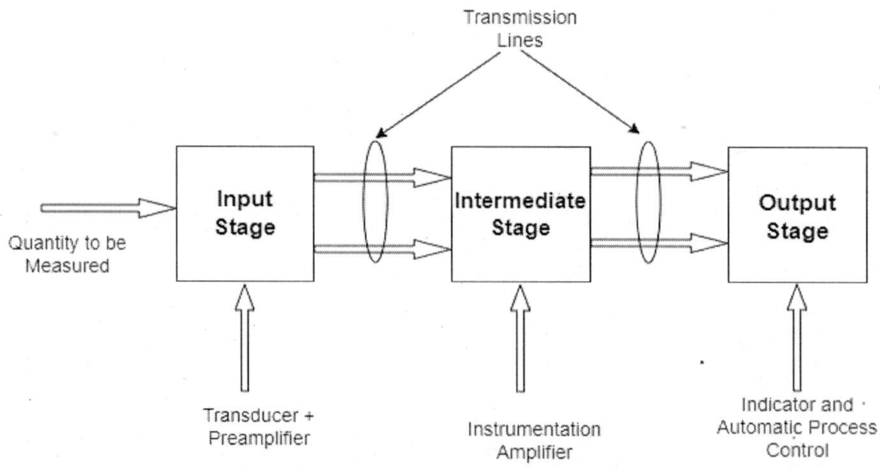

Fig.2.5 Instrumentation Amplifier

The transducer is a piece of equipment that can change the form of the energy that is being sent. For instance, when a strain gauge is subjected to pressure or force, there is a corresponding change in the amount of electrical energy that it produces. The output signal that is created by a transducer is measured by an instrumentation system, which is also frequently used to control the physical signal that is producing the output signal. A simplified version of such a system is depicted in the figure above. The input stage is made up of a pre-amplifier and a transducer of some kind; the specific type of transducer used is determined by the physical quantity that is going to be measured. In the output step, instruments such as metres, oscilloscopes, charts, or magnetic records could be utilised.

Transmission lines are depicted as the connecting lines between the blocks in Fig. 2.5. These connections are particularly useful when the transducer is located at a remote test site monitoring potentially dangerous conditions, such as high temperatures or liquid levels of combustible substances. Signals can be transferred from one unit to another via these transmission lines. The primary factors that determine the length of the transmission lines are the physical quantities that need to be monitored as well as the requirements of the system.

The output of the transducer is the source of the signal that is amplified by the instrumentation amplifier. Although certain transducers are capable of producing outputs that are strong enough to permit their use directly, the majority of them are not. The primary purpose of the instrumentation amplifier is to boost the relatively weak output signal from the transducer so that it can power the indicator or display; this is the amplifier's most important function. In a nutshell, the instrumentation amplifier is designed for precise, low-level signal amplification in situations where there is a requirement for accurate closed-loop gain in addition to low noise, low thermal drift, high input resistance, and low time drift. In addition, a high common-mode rejection ratio, a low power consumption, and a high slew rate are all desirable for greater performance.

Instrumentation operational amplifiers come in a wide variety; some examples include the LA 725, ICL7605, and LH0036. These amplifiers help to ensure that a circuit is exceptionally stable and accurate. These integrated circuits (ICs) are quite pricey due to the fact that they are extremely precise special-purpose circuits. The majority of the electrical parameters, such as offsets, drifts, and power consumption, are minimised while input resistance, CMRR, and supply range are optimised. Even in the form of modular components, certain instrumentation amplifiers can be tailored to meet the specific needs of individual installations.

It should come as no surprise that the specifications for instrumentation op-amps are significantly more stringent than those for general-purpose applications. On the other hand, the differential mode can be utilised with the general-purpose op-amp in situations when the requirements are not very stringent.

These amplifiers will hereafter be referred to as differential instrumentation amplifiers. In light of the fact that the vast majority of instrumentation systems incorporate a transducer into a bridge circuit, we will investigate a simpler differential instrumentation system arrangement that makes use of a transducer bridge circuit.

2.5 V to I Converter:

Fig. 2.6 shows a voltage to current converter in which load resistor R_L is floating (not connected to ground). The input voltage is applied to the non-inverting input terminal and the feedback voltage across R drives the inverting input terminal. This circuit is also called a current series negative feedback, amplifier because the feedback voltage across R depends on the output current i_L and is in series with the input difference voltage V_d. Writing the voltage equation for the input loop. $V_{in} = V_d + V_f$

But V_d »V_f since A is very large, therefore,

$$V_{in} = V_f \; V_{in} = RI_{in}$$

and since input current is zero.

$$I_{in} = V_{in} / R.$$

$$I_L = I_{in} = V_{in} ./ R$$

The value of load resistance does not appear in this equation. Therefore, the output current is independent of the value of load resistance. Thus the input voltage is converted into current; the source must be capable of supplying this load current.

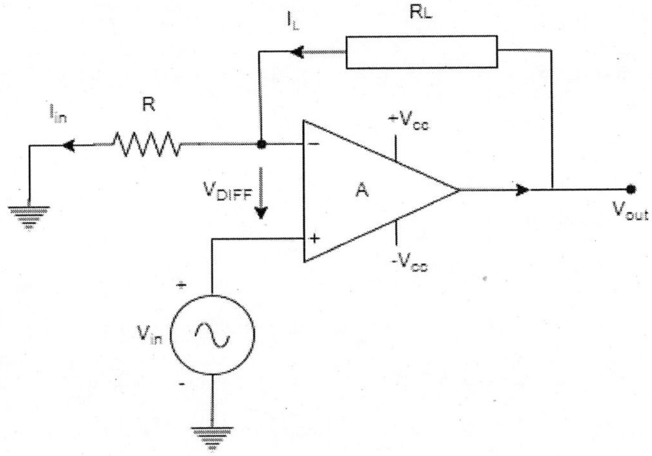

Fig.2.6: Circuit Diagram of V to I Converter

The maximum load current is V_{CC}/ R. In this circuit v_{in} may be positive or negative.

2.6 I to V Converter:

The circuit shown in fig 2.7 is a current to voltage converter.

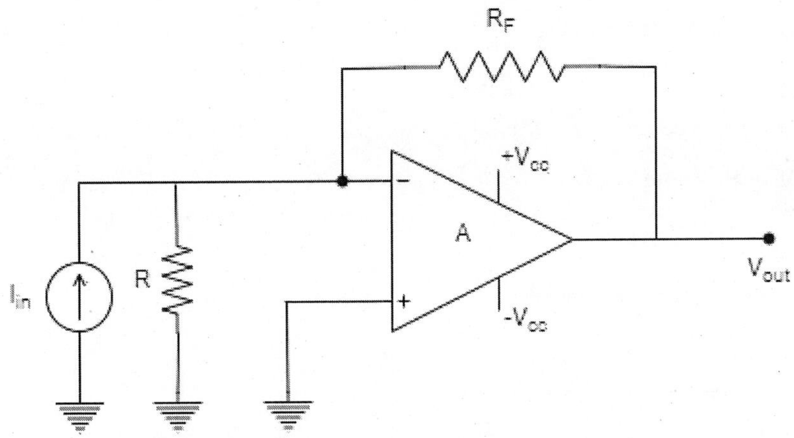

Fig.2.7 : Circuit Diagram of I to V Converter

Due to virtual ground the current through R is zero and the input current flows through R_f. Therefore, v_{out} =-R_f * i_{in}. The lower limit on current measure with this circuit is set by the bias current of the inverting input.

2.7 Sample and Hold circuits:

The sample and hold circuit, as its name implies samples an i/p signal and holds on to it last sampled value until the i/p is sampled again. Below fig2.8 shows a sample and hold circuit using an op-amp with an E- MOSFET. In this circuit the E-MOSFET works as a switch that is controlled by the sample and control voltage V_s, and the capacitor C serves as a storage element.

The analog signal V_{in} to be sampled is applied to the drain, and sample and hold control voltage V_s is applied to the gate of the E-MOSFET. During the positive portion of the V_s, the EMOSFET conducts and acts as a closed switch. This allows i/p voltage to charge capacitor C. In other words input voltage appears across C and in turn at the o/p as shown in fig 2.8. On the other hand, when V_s is zero, the EMOSFET is off and acts as open switch. The only discharge path for C is, through the op-amp. However the input resistance of the op-amp voltage follower is also very high; hence the voltage across C is retained.

The time periods T_s of the sample-and-hold control voltage V_s during which the voltage across the capacitor is equal to the input voltage are called sample periods. The time periods T_H of V_s during which the voltage across the capacitor is constant are called hold periods. The output of the op-amp is usually processed/ observed during hold periods. To obtain the close approximation of the input waveform, the frequency of the sample-and-hold control voltage must be significantly higher than that of the input.

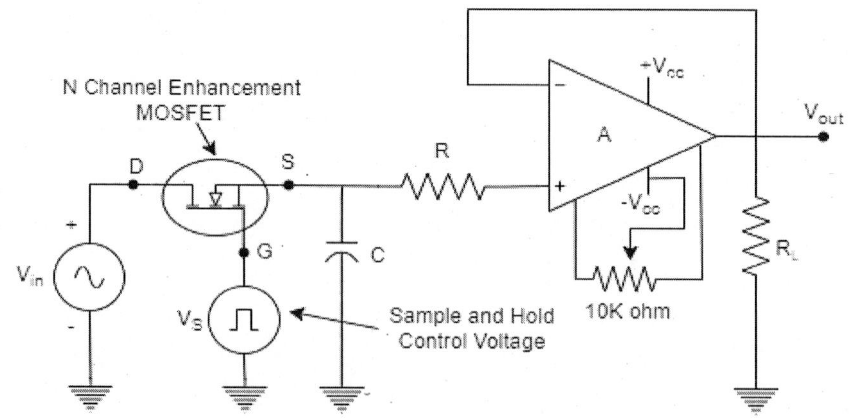

(a)

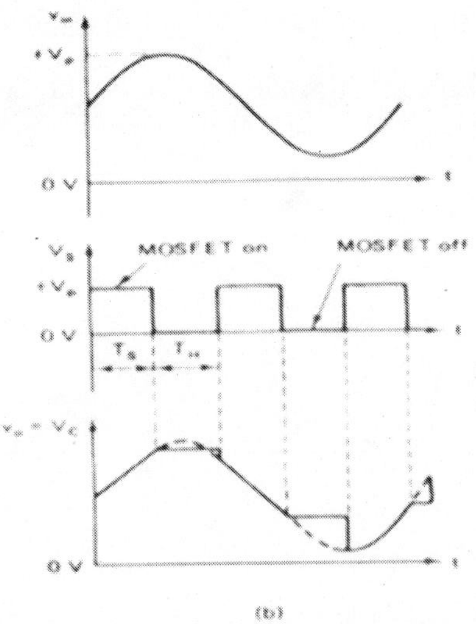

Fig.2.8.: a) sample and hold circuit b)input and output wave forms

2.8 Differentiator:

A circuit in which the output voltage waveform is the differentiation of input voltage is called differentiator as shown infig.2.9.

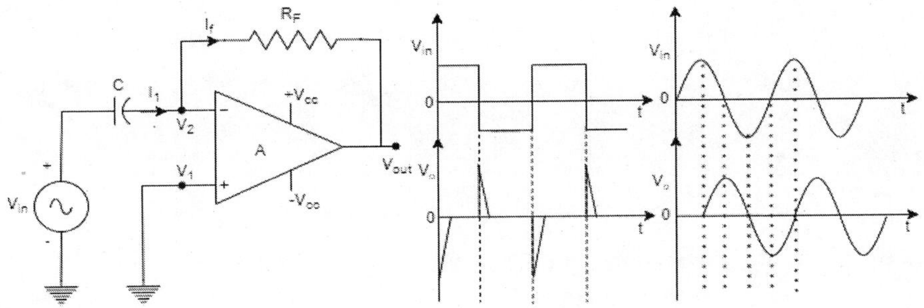

Fig.2.9: Circuit Diagram of Differentiator

The expression for the output voltage can be obtained from the Kirchoff's current equation written at node v2.

$$\text{Since,} \quad I_{in} = I_f$$

$$\text{Therefore,} \ C\frac{d}{dt}(V_{in} - 0) = \frac{0 - V_0}{R}$$

$$V_0 = -RC\frac{dV_{in}}{dt}$$

Thus the output v_o is equal to the RC times the negative instantaneous rate of change of the input voltage vin with time. A cosine wave input produces sine output. Fig.2.9 also shows the output waveform for different input voltages.

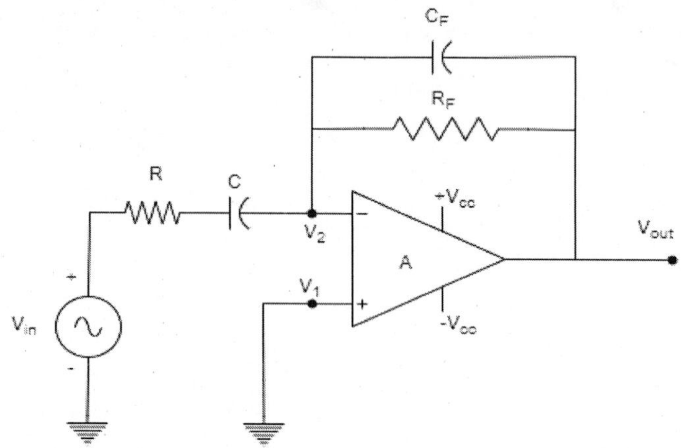

Fig.2.10: Circuit Diagram of Differentiator

The input signal will be differentiated properly if the time period T of the input signal is larger than or equal to $R_f C$. As the frequency changes, the gain changes. Also at higher frequencies the circuit is highly susceptible at high frequency noise and noise gets amplified. Both the high frequency noise and problem can be corrected by adding, few components. as shown in fig.2.10.

2.9 Integrator:

A circuit in which the output voltage waveform is the integral of the input voltage waveform is called integrator. Fig.2.11 shows an integrator circuit using OPAMP.

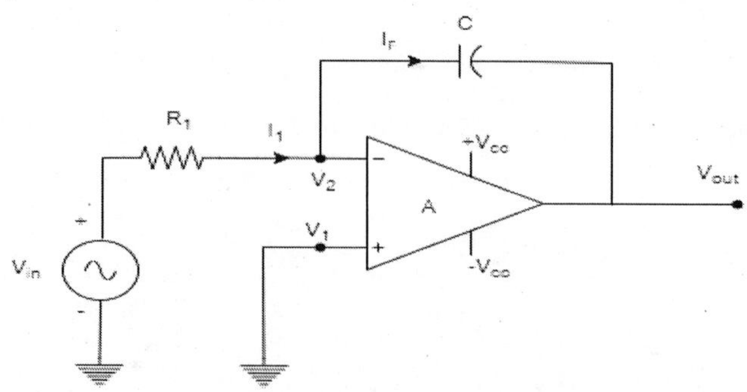

Fig.2.11: Circuit Diagram of Integrator

Here, the feedback element is a capacitor. The current drawn by OPAMP is zero and also the V_2 is virtually grounded.Therefore, $I_1 = I_f$ and $V_2 = V_1 = 0$

$$\frac{V_{in} - 0}{R} = C\frac{d(0 - V_0)}{dt}$$

Integrating both sides with respect to time from 0 to t, we get

$$\int_0^t \frac{V_{in}}{R} dt = \int_0^t C\left(\frac{d(-V_0)}{dt}\right) dt$$

$$= C(-V_0) + V_0/_{t=0}$$

$$if\ V_0/_{t=0} = 0V, then$$

$$V_0 = -\frac{1}{RC}\int_0^t V_{in}\, dt$$

The output voltage is directly proportional to the negative integral of the input voltage and inversely proportional to the time constant RC. If the input is a sine wave the output will be cosine wave. If the input is a square wave, the output will be a triangular wave. For accurate integration, the time period of the input signal T must be longer than or equal to RC.

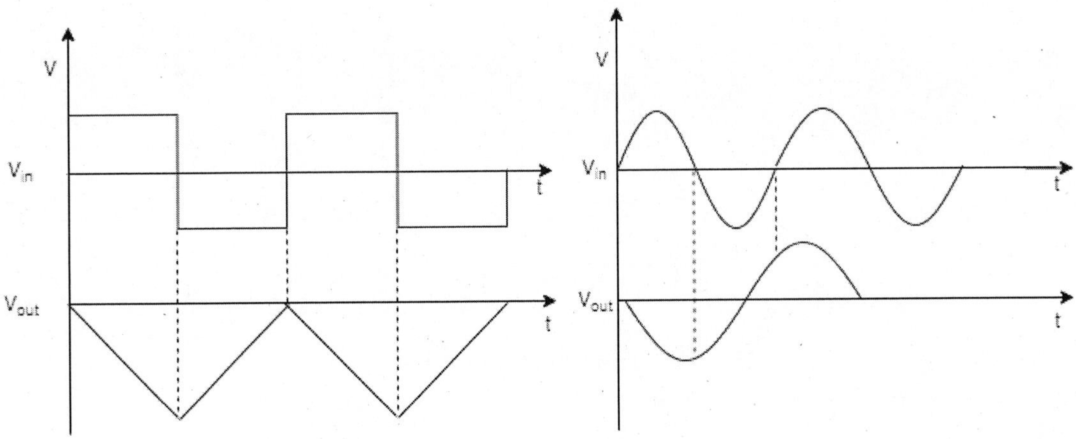

Fig.2.12 : Input and Output wave forms

2.10 comparator:

2.10.1Voltage comparator circuit:

Voltage comparator is a circuit which compares two voltages and switches the output to either high or low state depending upon which voltage is higher. A voltage comparator based on opamp is shown here. Fig.2.13 shows a voltage comparator in inverting mode and Fig shows a voltage comparator in non-inverting mode.

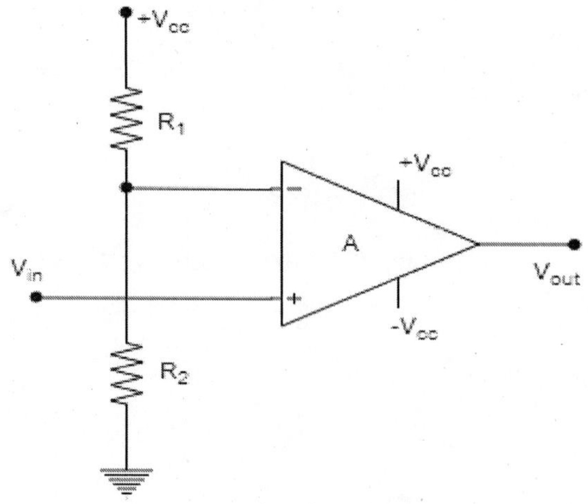

Non Inverting Comparator

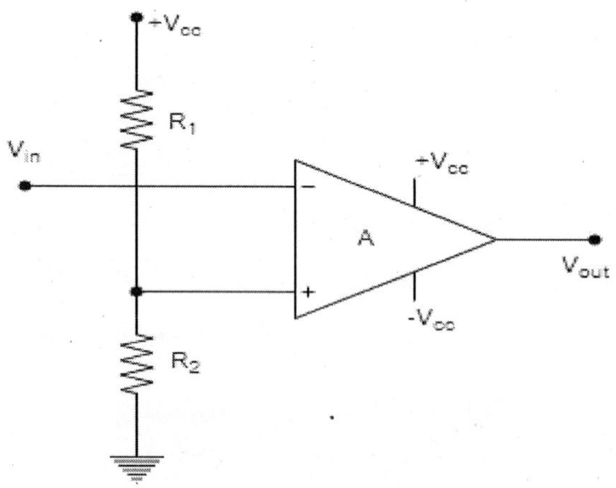

Inverting Comparator

Fig.2.13 : Circuit Diagram of Comparator

2.10.2 Non inverting comparator:

In non-inverting comparator the reference voltage is applied to the inverting input and the voltage to be compared is applied to the non-inverting input. Whenever the voltage to be compared (V_{in}) goes above the reference voltage, the output of the opamp swings to positive saturation (V+) and vice versa. Actually what happens is that, the difference between V_{in} and V_{ref}, ($V_{in} - V_{ref}$) will be a positive value and is amplified to infinity by the opamp. Since there is no feedback resistor Rf, the opamp is in open loop mode and so the voltage gain (A_v) will be close to infinity. So the output voltage swings to the maximum possible value ie; V+. Remember the equation $A_v = 1 + (R_f/R_1)$. When the Vin goes below V_{ref}, the reverse occurs.

2.10.3 Inverting comparator.

In the case of an inverting comparator, the reference voltage is applied to the non-inverting input and voltage to be compared is applied to the inverting input. Whenever the input voltage (V_{in}) goes above the V_{ref}, the output of the op-amp swings to negative saturation. Here the difference between two voltages (V_{in}-V_{ref}) is inverted and amplified to infinity by the op-amp. Remember the equation $A_v = -R_f/R_1$. The equation for voltage gain in the inverting mode is $Av = -R_f/R_1$.Since there is no feedback resistor, the gain will be close to infinity and the output voltage will be as negative as possible i.e., V-.

2.10.4 Practical voltage comparator circuit.

A practical non inverting comparator based on uA741 opamp is shown below. Here the reference voltage is set using the voltage divider network comprising of R_1 and R_2. The equation is $V_{ref} = (V+/ (R_1 + R_2)) \times R_2$. Substituting the values given in the circuit diagram into this equation gives $V_{ref} = 6V$. Whenever Vin goes above 6V, the output swings to ~+12V DC and vice versa. The circuit is powered from a +/- 12V DC dual supply.

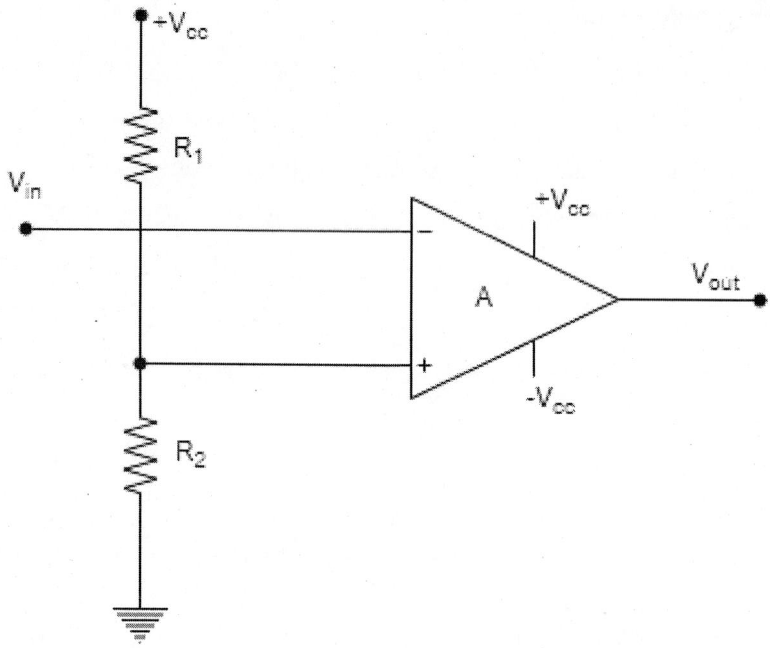

Fig.2.14 : Circuit diagram of Practical voltage comparator.

2.11 Schmitt Trigger:

Below fig.2.15 shows an inverting comparator with +ve feedback. This circuit converts an irregular shaped wave forms to a square wave form or pulse. The circuit is known as schmitt trigger or squaring circuit. The input voltage being triggers the o/p Vo every time it exceeds certain voltage levels called the upper threshold voltage V_{ut} and lower threshold voltage V_{lt}.In fig 2.15 these threshold voltages are obtained by using the voltage divider R_1, R_2, where the voltage across R_1 is feedback to +ve input. The voltage across R_1 is a variable reference, threshold voltage that depends on the value and polarity of the output voltage Vo. when V_o=+V_{sat}, the voltage across R_1 is called the upper threshold voltage V_{ut}.

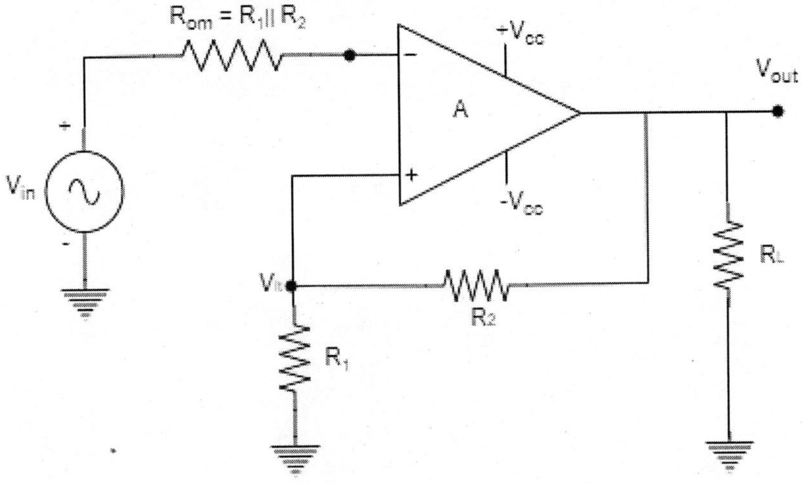

Fig 2.15 Schmitt Trigger

The input voltage V_{in} must be slightly more positive then V_{ut} in order to cause the out put V_o to switch from $+V_{sat}$ to $-V_{sat}$.as long as V_{in} less than V_{ut}, V_o is at $+V_{sat}$. using the voltage divider rule, On the other hand, when V_o=-V_{sat}, the voltage across R_1 is referred to as the lower threshold voltage,V_{lt}. Vin must be slightly more negative than V_{lt}.in order to cause V_o to switch from-V_{sat} to +Vsat.in other words, for Vin values greater than V_{lt}, V_o is at – V_{sat}. Thus if the threshold voltages V_{ut} and V_{lt} are made large than the input noise voltages, the positive fed back will eliminate the false output transitions. Also the positive feedback because of its regenerative action will make V_o switch faster between $+V_{sat}$ and – V_{sat}.

2.12 First-order low-pass butter worth filter

Fig. 2.16 shows a first-order low-pass Butterworth filter that uses an RC network for filtering. Note that the op- amp is used in the non-inverting configuration; hence it does not load down the RC network. ResistorsR_1 and R_F determine the gain of the filter. According to the voltage-divider rule, the voltage at the non- inverting terminal (across capacitor C) is

$$V_1 = \frac{-jX_C}{R - jX_C} V_{in}$$

$$j = \sqrt{-1}\ and - jX_C = \frac{1}{j2\pi fC}$$

$$V_1 = \frac{V_{in}}{1 + j2\pi fRC}$$

$$V_0 = 1 + \frac{R_F}{R_1} V_1$$

$$V_0 = \left(1 + \frac{R_F}{R_1}\right) \frac{V_{in}}{1 + j2\pi fRC}$$

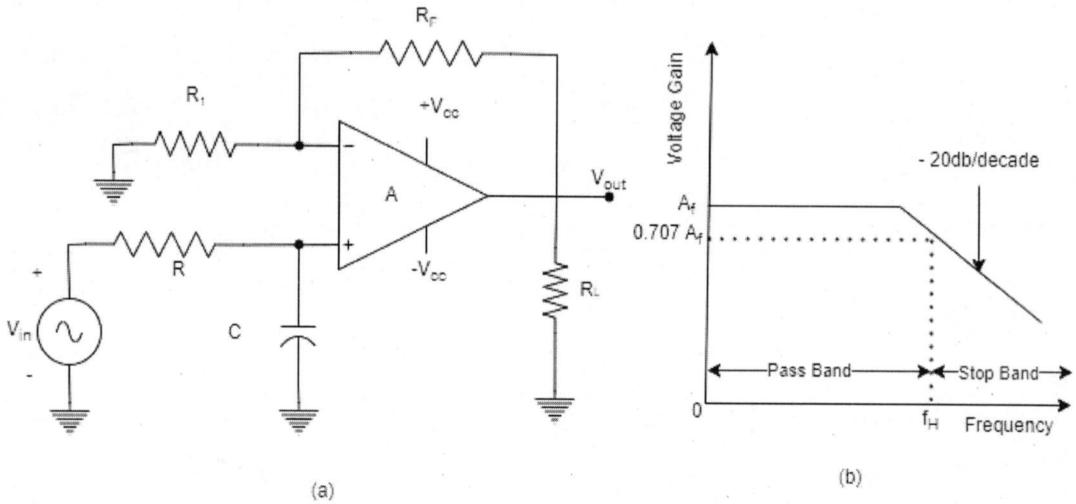

Fig.2.16 : First order Low Pass Butter Worth Filter (a) circuit (b) Response

Where V_o/V_{in} = gain of the filter as a function of frequency. $A_F = \left(1+\frac{R_F}{R_1}\right)$ = pass band gain of the filter .f = input frequency of the filter and $f_H = \frac{1}{2\pi RC}$ = upper cut-off frequency of the filter.The gain magnitude and phase angle equations of the low-pass filter can be obtained by converting Equation into its equivalent polar form, as follows:

$$\left|\frac{V_0}{V_{in}}\right| = \frac{A_F}{\sqrt{1+\left(\frac{f}{f_H}\right)^2}}$$

$$\Phi = -\tan^{-1}\left(\frac{f}{f_H}\right)$$

Where φ is the phase angle in degrees. The operation of the low pass filter can be verified from the gain magnitude equation:

1. At very low frequencies that is , $f<f_H$, $\left|\frac{V_0}{V_{in}}\right| = A_F$

2. At $f=f_H$, $\left|\frac{V_0}{V_{in}}\right| = \frac{A_F}{\sqrt{2}} = 0.707A_F$

3. At $f>f_H$, $\left|\frac{V_0}{V_{in}}\right| < A_F$

A low-pass filter can be designed by implementing the following steps:

1. Choose a value of high cutoff frequency f_H.

2. Select a value of C less than or equal to 1 μF. Mylar or tantalum capacitors are recommended for better performance.

3. Calculate the value of R using R = $\left(\frac{1}{2\pi f_H C}\right)$

4. Finally, select values of R_1 and R_F dependent on the desired pass band gain A_F using

$$A_F = 1 + \frac{R_F}{R_1}$$

Once a filter designed; there may sometimes be a need to change its cut-off frequency. The procedure used to convert an original cut-off frequency f_H to a new cut-off frequency f'_H is called frequency scaling. Frequency scaling is accomplished as follows. To change a high cutoff frequency, multiple R or C, but not both, by the ratio of the original cutoff frequency to the new cutoff frequency.

2.12.1 Second-order Low-pass Butter worth Filter

A stop-band response having a 40-dB/decade roll-off is obtained with the second order low- pass filter. A first-order low-pass filter can be converted into a second order type simply by using an additional RC network, as shown in Fig.2.17.

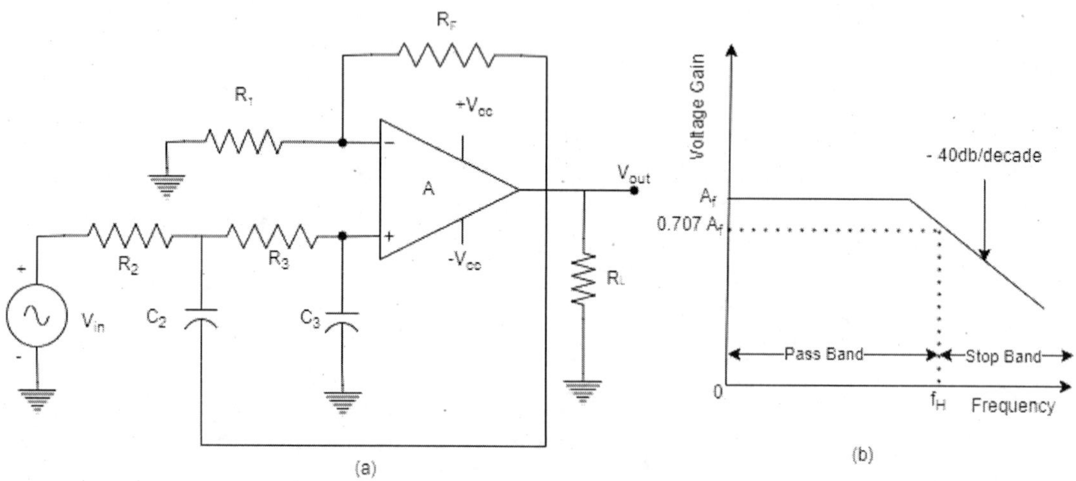

Fig.2.17 : Second order Low Pass Butter Worth Filter (a)Circuit(b)Frequency Response

Second-order filters are important because higher-order filters can be designed using them. The gain of the second-order filter is set by R_1 and R_F, while the high cutoff frequency f_H is determined by R_2, C_2, R_3, and C_3, as follows

$$f_H = \frac{1}{2\pi\sqrt{R_2 R_3 C_2 C_3}}$$

Furthermore, for a second-order low-pass Butterworth response, the voltage gain magnitude equation is

$$\left|\frac{V_0}{V_{in}}\right| = \frac{A_F}{\sqrt{1+\left(\frac{f}{f_H}\right)^4}}$$

Where Vo/Vin = gain of the filter as a function of frequency, $A_F = \left(1 + \frac{R_F}{R_1}\right)$ = pass band gain of the filter, f = input frequency of the filter and $f_H = \frac{1}{2\pi\sqrt{R_2R_3C_2C_3}}$ = upper cut-off frequency of the filter

2.13 High pass filter

2.13.1 First-order high-pass butter worth filter

High-pass filters are often formed simply by interchanging frequency-determining resistors and capacitors in low-pass filters. That is, a first-order high-pass filter is formed from a first-order low-pass type by interchanging components Rand C.

Similarly, a second-order high-pass filter is obtained from a second-order low-pass filter if R and Care interchanged, and so on. Figure shows a first-order high-pass Butterworth filter with a low cutoff frequency of f_L. This is the frequency at which the magnitude of the gain is 0.707 times its pass band value. Obviously, all frequencies higher than f_L are pass-band frequencies, with the highest frequency determined by the closed- loop bandwidth of the op-amp.

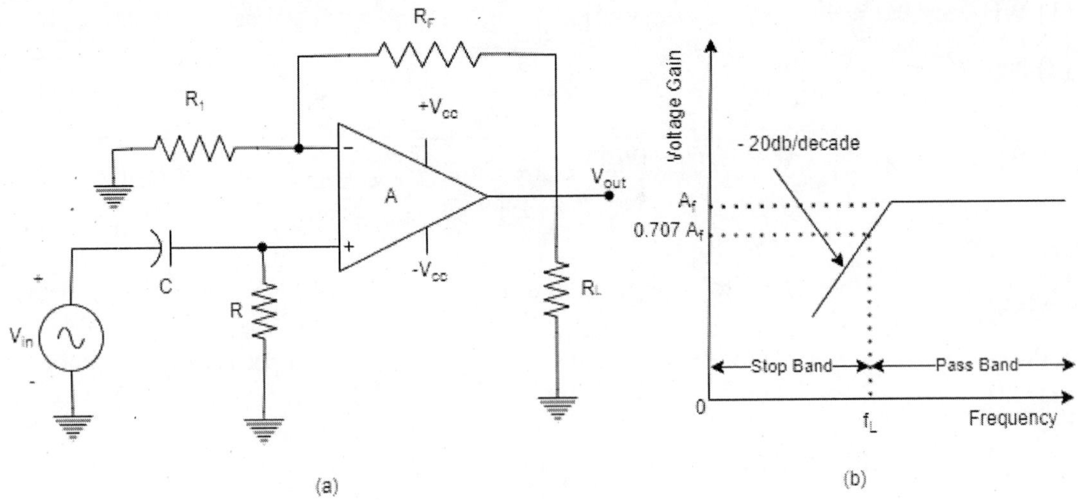

Fig.2.18 : (a) First order High Pass Butter worth Filter (b) Frequency Response For the first-order high-pass filter of Figure (a), the output voltage is

$$V_0 = \left(1 + \frac{R_f}{R_1}\right) \frac{j2\pi fRC}{1 + j2\pi fRC} V_{in}$$

$$\frac{V_0}{V_{in}} = A_F \frac{j\left(\frac{f}{f_L}\right)}{\sqrt{1 + j\left(\frac{f}{f_L}\right)}}$$

Hence the magnitude of the voltage gain is

$$\left|\frac{V_0}{V_{in}}\right| = A_F \frac{\left(\frac{f}{f_L}\right)}{\sqrt{1 + \left(\frac{f}{f_L}\right)^2}}$$

2.14 Band-Pass Filters

A band-pass filter has a pass band between two cut-off frequencies f_H and f_L such that $f_H > f_L$. Any input frequency outside this pass band is attenuated. Basically, there are two types of band-pass filters:

(1) Wide band pass, and

(2) Narrow band pass.

Unfortunately, there is no set dividing line between the two. However, we will define a filter as wideband pass if its figure of merit or quality factor $Q<10$.On the other hand, if we will call the filter a narrow band-pass filter. Thus Q is a measure of selectivity, meaning the higher the value Q, the more selective is the filter or the narrower its bandwidth (BW). The relationship between Q, the3- dB bandwidth, and the center frequency fc is given by For the wideband-pass filter the center frequency f_c can be defined as where f_H =high cut off frequency(Hz) f_L=low cut off frequency of the wideband-pass filter (Hz) In a narrowband- pass filter, the output voltage peaks at the center frequency.

2.14.1 Wide-band pass filter

A wide band-pass filter can be formed by simply cascading high-pass and low-pass sections and is generally the choice for simplicity of design and performance. To obtain ±20dB/decade band-pass, first-order high pass and first order low-pass sections are cascaded; fora±40-dB/decade band-pass filter, second-order high- pass and second- order low-pass sections are connected in series. Figure showsthe±20-dB/decade wideband pass filter, which is composed of first-order high- pass and first-order low-pass filters. To realize a band- pass response, however, f_H must be larger than f_L.

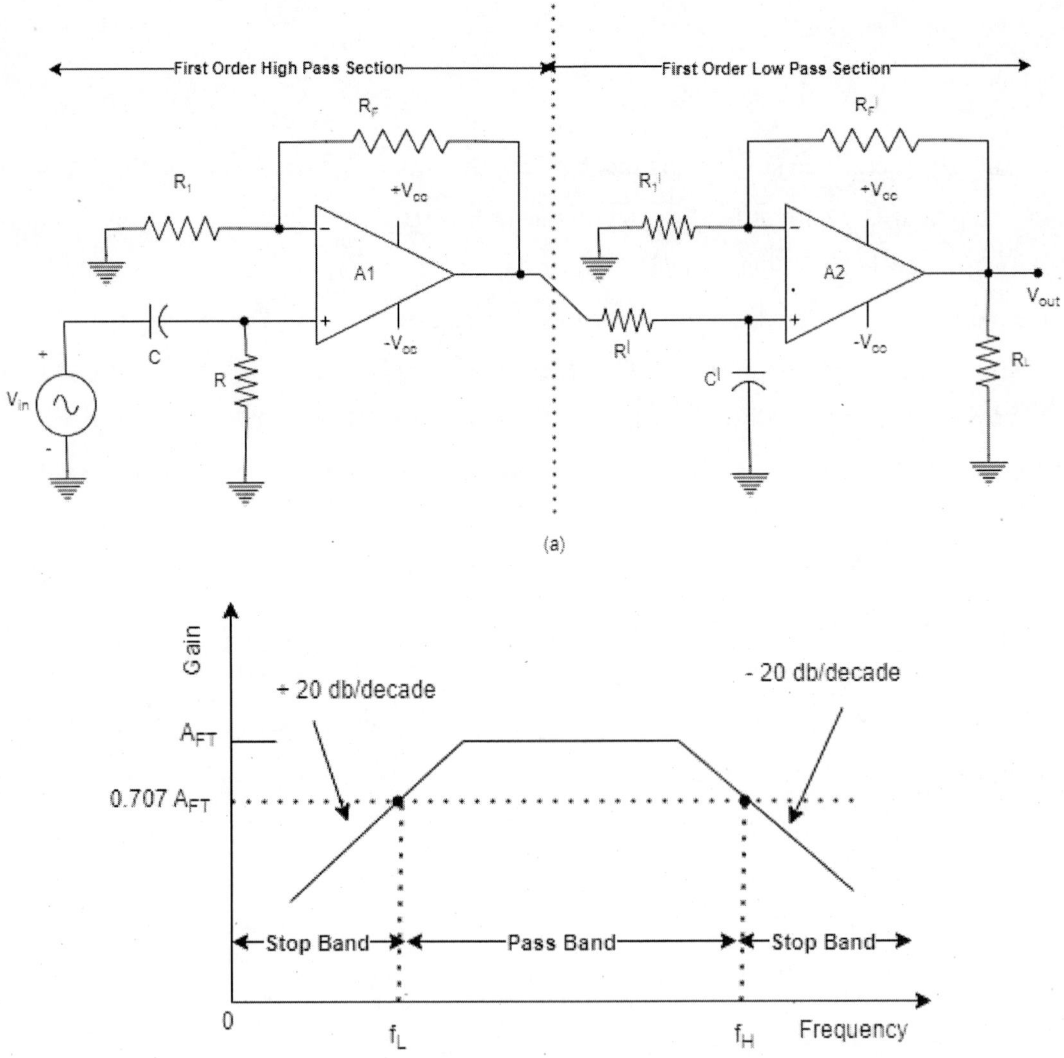

Fig .2.19(a)±20dB/decade Wide Band Pass Filter (b)Frequency Response

Since the band-pass gain is 4, the gain of the high-pass as well as low-pass sections could be set equal to 2. That is, input and feedback resistors must be equal in value, say10 kΩ each. The complete band-pass filter is shown in Fig 2.19 (a) & (b). The voltage gain magnitude of the band-pass filter is equal to the product of the voltage gain magnitudes of the high-pass and low-pass filters.

$$\left|\frac{V_O}{V_{in}}\right| = \frac{A_F\left(\frac{f}{f_L}\right)}{\sqrt{1+\left(\frac{f}{f_L}\right)^2}}$$

$$\left|\frac{V_O}{V_{in}}\right| = \frac{A_{FT}\left(\frac{f}{f_L}\right)}{\sqrt{\left[1+\left(\frac{f}{f_L}\right)^2\right]\left[1+\left(\frac{f}{f_H}\right)^2\right]}}$$

Where A_{FT} =total pass band gain

f= frequency of the input signal (Hz)

f_L=low cut off frequency(Hz)

f_H=high cut off frequency(Hz)

2.14.2 Narrow Band-Pass Filter

The narrow band-pass filter using multiple feedback is shown in Figure2.20. As shown in this figure, the filter uses only one op-amp. Compared to all the filters discussed so far, this filter is unique in the following respects:

1. It has two feedback paths, hence the name multiple-feedback filter.
2. The op-amp is used in the inverting mode.

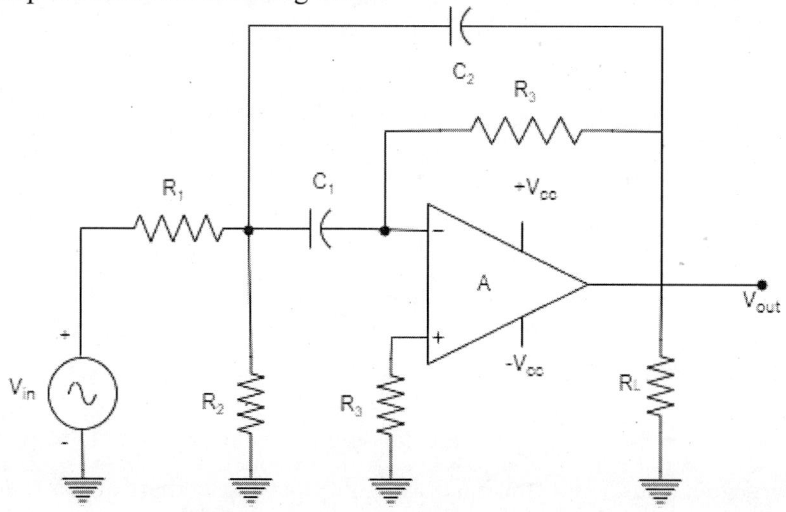

Fig.2.20: Multiple Feedback Narrow Band Pass Filter

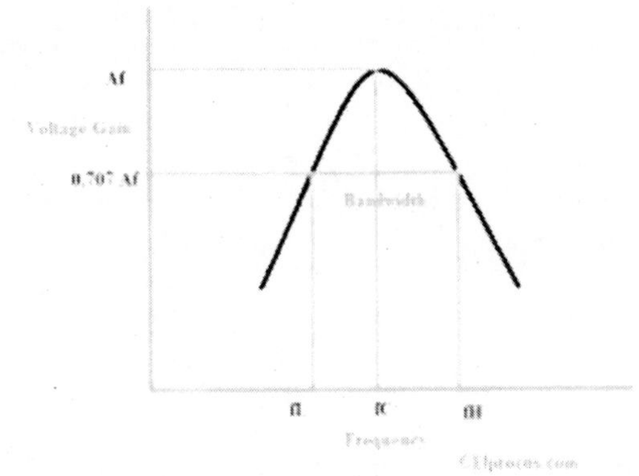

Fig.2.20 : (b) Frequency Response

Generally, the narrow band-pass filter is designed for specific values of center frequency *fc* and Q or *fc* and bandwidth. The circuit components are determined from the following relationships. To simplify the design calculations, choose $C_1 = C_2 = C$.

$$R_1 = \frac{Q}{2\pi f_c C A_F}$$

$$R_2 = \frac{Q}{2\pi f_c C(2Q^2 - A_F)}$$

$$R_3 = \frac{Q}{\pi f_c C}$$

Where A_F is the gain at f_c, given by, $A_F = \frac{R_3}{2R_1}$, The gain A_F, however, must satisfy the condition $A_F = 2Q^2$. Another advantage of the multiple feedback filter of Figure is that its center frequency *fc* can be changed to a new frequency *f'c* without changing the gain or bandwidth. This is accomplished simply by changing R_2 to R'_2 so that

$$R'_2 = R_2 \left(\frac{f_c}{f'_c}\right)^2$$

2.15 Band-Reject Filters

The band-reject filter is also called a band-stop or band-elimination filter. In this filter, frequencies are attenuated in the stop band while they are passed outside this band.

As with band-pass filters, the band-reject filters can also be classified as wideband- reject and narrowband-reject. The narrow band-reject filter is commonly called the notch filter. Because of its higher Q (>10), the bandwidth of the narrow band-reject filter is much smaller than that of the wideband- reject filter.

2.15.1 Wide Band-Reject Filter

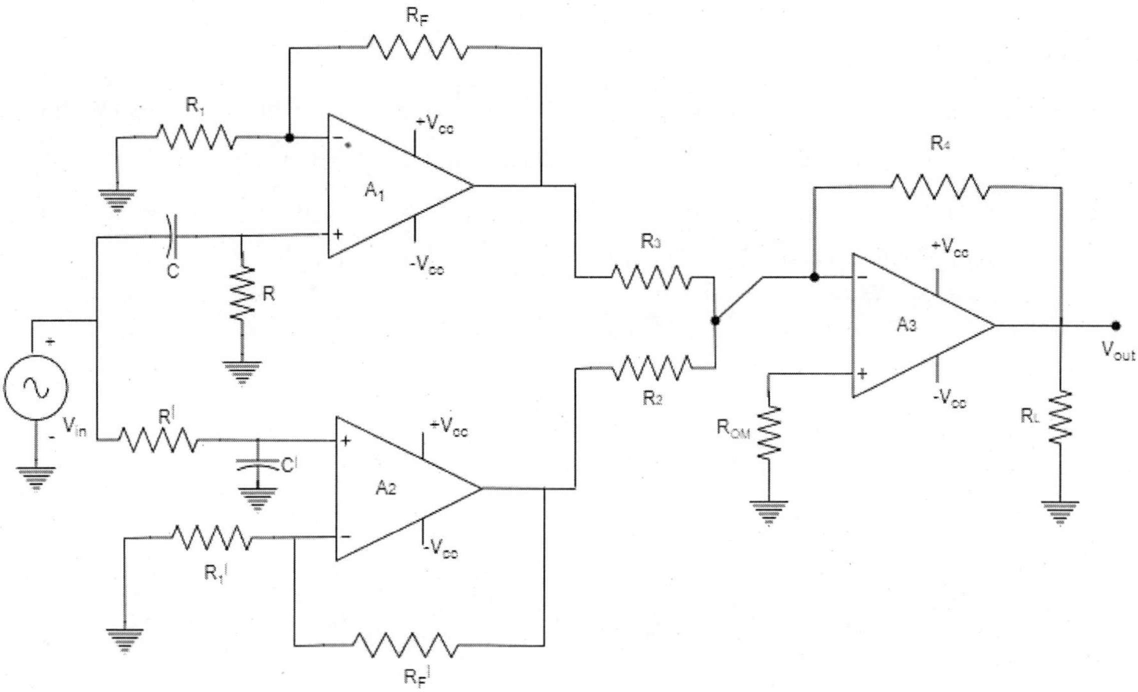

Fig.2.21 (a).Wide Band Reject Filter

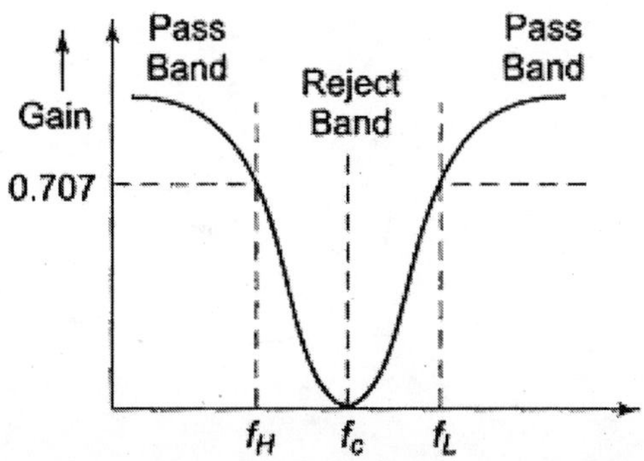

Fig .2.21(b) Frequency Response

Figure 2.21(a) shows a wide band-reject filter using a low-pass filter, a high-pass filter, and assuming amplifier. To realize a band-reject response, the low cut off frequency f_L of the high-pass filter must be larger than the high cut off frequency f_H of the low-pass filter. In addition, the pass band gain of both the high-pass and low-pass sections must be equal. The frequency response of the wideband-reject filter is shown in Fig 2.21(b).

2.16 All-Pass Filter

As the name suggests, an all-pass filter passes all frequency components of the input signal without attenuation, while providing predictable phase shifts for different signals. When signals are transmitted over transmission lines, such as telephone wires, they undergo change in phase. To compensate for these phase changes, all-pass filters are required.

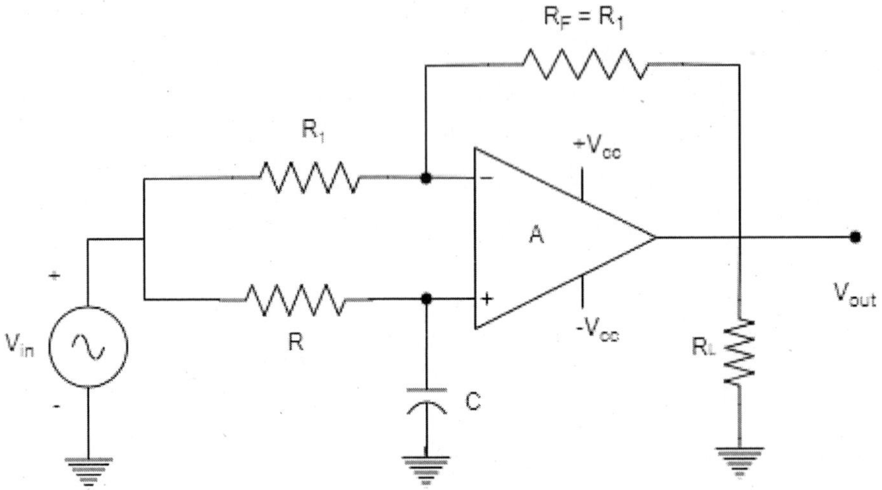

Fig.2.22 (a)All Pass Filter

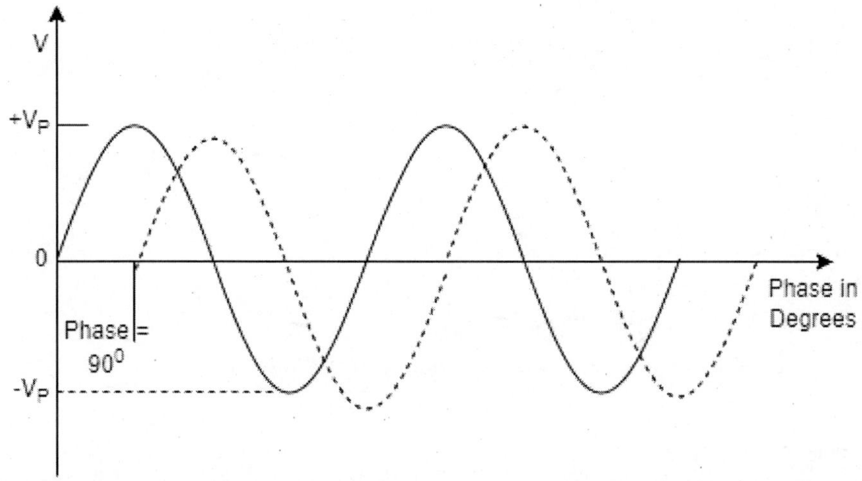

Fig.2.22 (b) Phase Shift between Input And Output frequencies of the input signal.

The all- pass filters are also called delay equalizers or phase correctors. Figure2.22 (a) shows an all-pass filter where in $R_F = R_1$. The output voltage Vo of the filter can be obtained by using the superposition theorem:

But -j =1/j and $X_C = 1/2\Pi fC$. Therefore, substituting for X_C and simplifying, we get

$$V_0 = V_{in}\left(-1 + \frac{2\cdot}{j2\pi fRC + 1}\right)$$

$$\frac{V_0}{V_{in}} = \frac{1 - j2\pi fRC}{1 + j2\pi fRC}$$

Where f is the frequency of the input signal in hertz. Equation indicates that the amplitude of V_o/V_{in} is unity; that is, $|V_o| = |V_{in}|$ throughout the useful frequency range, and the phase shift between V_o and V_{in} is a function of input frequency f. The phase angle φ is given by

$$\Phi = -2\tan^{-1}\left(\frac{2\pi fRC}{1}\right)$$

Where φ is in degrees, in hertz, R in ohms, and C in farads. Equation is used to find the phase angle φ if f, R, and C are known. Figure2.22 (b) shows a phase shift of 90° between the input V_{in} and output V_o. That is, V_o lags V_{in} by90°. For fixed values of R and C, the phase angle φ changes from 0to 180°asthe frequency f is varied from 0 to ∞. In Figure 2.22(a), if the positions of R and

C are interchanged, the phase shift between input and output becomes positive. That is, output V_o leads input V_{in}.

UNIT SUMMARY

- An op-amp is a high gain and direct coupled amplifier, and the voltage gain can be controlled by the externally connected feedback components.
- The op-amp can be used in amplifier and signal processing applications involving dc to several MHz of frequency ranges.
- Sign changer circuit is called phase inverter.
- Phase shift circuits are constant-delay filters or all-pass filters.
- A current to voltage converter is called transresistance amplifier
- The instrumentation amplifier is used for precise amplification of low level output signals where low noise, low thermal and time drifts, high input impedance and accurate closed-loop gains are required.
- They achieve high CMRR, high gain stability with low temperature coefficient, low dc offset and low output impedance. AD521, AD524 and AD624 manufactured by Analog Devices, and mA725, ICL7605 , and LH0036 are some of the commercial instrumentation amplifiers.
- AC amplifiers can be realised in two configurations, namely, (i) inverting ac amplifier, and (ii) noninverting ac amplifier.
- Integrator or integrating amplifier is a circuit in which the output voltage waveform is the time integral of the input voltage waveform
- The differentiators can be used as waveshaping circuits and edge detectors in FM demodulators.
- The logarithmic amplif ier is a current to voltage converter
- Antilog amplifier is a decoding circuit which converts the logarithmically encoded signal back to the original signal

Multiple Choice Questions

1. A logarithmic amplifier can be used to do the following function

    (a) Addition (b) Subtraction

    (b) Noise rejection (d) Good selectivity

[Ans. (a)]

2. The output waveform of an op amp integrator circuit with square-wave input signal is

    (a) Triangular waveform (b) Ramp

    (b) Saw-tooth wave (d) Spikes

[Ans. (a)]

3. The output waveform of an op amp differentiator circuit with square-wave input signal is
   (a) Triangular waveform (b) Ramp
   (b) Saw-tooth wave (d) Spikes

[Ans. (d)]

4. Operational amplifier has got its name from the following mathematical operations
   (a) Addition (b) Integration
   (b) Differentiation (d) All of the above

[Ans. (d)]

5. Filter that has maximally flat response in the pass Band is
   (a) Chebyshev (b) Butterworth
   (b) All-pass filter (d) Switched capacitor filter

[Ans. (b)]

6. Gain of filter circuits has following measuring unit
   (a) Bel (b) dB (c) Neper (d) degrees

[Ans. (b)]

7. The frequency response of second-order LPF attenuates at the rate of
   (a) −10 dB/decade (b) −20 dB/decade
   (b) −30 dB/decade (d) −40 dB/decade

[Ans. (d)]

Short and Long Answer Type Questions

1. Explain how a basic inverting amplifier configuration of an op-amp is used as the following circuits
   a) Sign changer,
   b) Scale changer,
   c) Phase shift circuit
2. With a circuit diagram explain the working of a current to voltage converter.
3. What is the instrumentation amplifier? Draw a system whose gain is controlled by a variable resistance?
4. What is the basic function of a differentiator?
5. Draw a circuit for a first order active high-pass filter.
6. Define an all-pass filter. How can it be justifiably called as phase shift circuits?

REFERENCES AND SUGGESTED READINGS

1. Coughlin, R. F. and Frederick F. Driscoll (2001). Operational Amplifiers and Linear Integrated Circuits, Sixth Edition. Upper Saddle River, NJ: Prentice Hall.
2. Dailey, Denton J. (1989) Operational Amplifiers and Linear Integrated Circuits: Theory and Applications. New York: The McGraw-Hill Companies.
3. Franco, Sergio (2002). Design with Operational Amplifiers and Analog Integrated Circuits. New York: The McGraw-Hill Companies.
4. Gayakwad, Ramakant A. (1999). Op-Amps and Linear Integrated Circuits. Fourth edition. Upper Saddle River, NJ: Prentice Hall Inc.
5. Jacob, J. Michael (1996) Applications and Design with Analog Integrated Circuits. Upper Saddle River, NJ: Prentice Hall Inc.
6. *Ramakanth A. Gayakwad* (2000),Op-Amps and Linear Integrated Circuits -, *Prentice-Hall, 4th edition.*
7. S.Salivahanan,V S Kanchana Bhaskaran(2015),Linear Integrated Circuits, Second Edition, Tata McGrawa Hill Edition.
8. B.Visweswara Rao(2015), Linear Integrated Circuits, First Edition, Pearson India Education Services Pvt. Ltd

Dynamic QR Code for Further Reading

3 Analog Multiplier and PLL

UNIT SPECIFICS

Through this unit we have discussed the following aspects:

- *Different types of Multiplier Circuits;*
- *Different steps in the operation of Phase Locked Loop;*
- *Operation of Voltage Controlled Oscillator;*
- *Discussion on different applications of PLL;*

RATIONALE

In the design of instrumentation, communication, and control systems, it is frequently necessary to perform non-linear operations on analogue signals. The processes of signal creation, rectification, modulation, and demodulation, frequency translation, multiplication, and division are all included in these operations. This chapter will explain the methods that are utilised the most frequently as well as the integrated circuit components that are necessary to carry out these processes.

First, we will talk about the Gilbert multiplier cell, which is the foundation of a broad variety of other circuits. Next, we will talk about how the Gilbert multiplier cell may be used to create a complete four-quadrant multiplier. There is also an explanation of the four-quadrant multiplier circuit, which is based on the variable transconductance technique. After that comes the discussion of the analogue multiplier IC AD633, which is then followed by the significant uses of the multiplier ICs.

Next part of the chapter deals with Phase Locked Loops and its applications in the field of communication systems.

PRE-REQUISITES

Electronic devices and circuits, Circuit theory

UNIT OUTCOMES

List of outcomes of this unit is as follows:

U3-O1: Describe multiplier Circuit

U3-O2: Explain the different multiplication techniques

U3-O3: Describe the operation of PLL

U3-O4: Discuss about various applications of PLL

U3-O5: Explain the operation of Voltage controlled oscillator

Unit-3 Outcomes	**EXPECTED MAPPING WITH COURSE OUTCOMES** (*1-Weak Correlation; 2- Medium Correlation; 3-Strong Correlation*)					
	CO-1	**CO-2**	**CO-3**	**CO-4**	**CO-5**	**CO-6**
U3-01	2	3	2	1	2	3
U3-02	1	3	1	2	1	2
U3-03	2	3	1	2	2	3
U3-04	1	3	2	1	1	2
U3-05	-	3	1	1	1	3

3.1 Analog Multipliers:

An output from a multiplier, denoted by V0, has a value that is proportional to the product of the two inputs, denoted by V_x and V_y. In other words, V_0 equals K times V_x times V_y, where K is the scaling factor that is typically kept at (1/10) V^{-1}. When it comes to analogue multiplication, there are a few different approaches that can be taken.

3.2 A simple multiplier using an Emitter coupled Transistor pair:

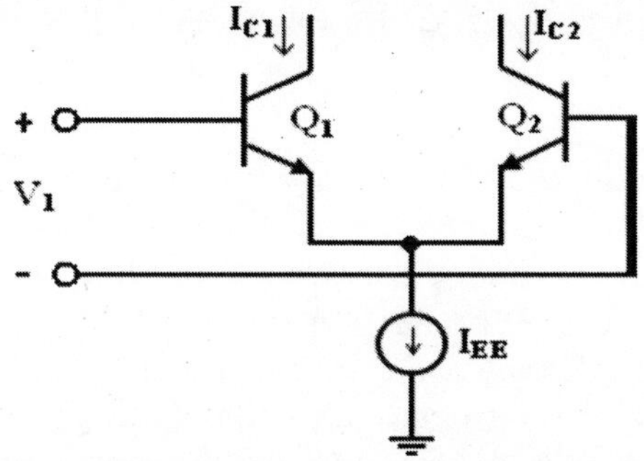

Fig.3.1.Emitter Coupled Transistor Pair

The illustration in Figure3.1 depicts a circuit that makes use of an emitter coupled pair. The differential input voltage is related to the output currents I_{C1} and I_{C2}, which can be written as:

$$I_{C1} = \frac{I_{EE}}{1 + e^{-V_1/V_T}}$$

$$I_{C2} = \frac{I_{EE}}{1 + e^{V_1/V_T}}$$

Where V_T is thermal voltage. The difference between two output currents is given by,

$$\Delta I_C = I_{C1} - I_{C2} = I_{EE} \tanh\left(\frac{V_1}{2V_T}\right)$$

Figure illustrates the dc transfer characteristics that are shown by the emitter-coupled pair.

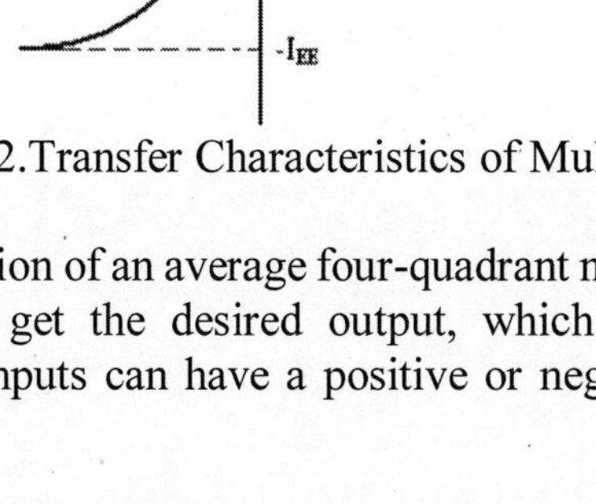

Fig.3.2.Transfer Characteristics of Multiplier

The characteristics of the transmission of an average four-quadrant multiplier are depicted in figure below in Figure3.2. In order to get the desired output, which is displayed in the transfer characteristics, either one of the inputs can have a positive or negative value. It shows that the emitter coupled pair can be used as a simple multiplier using this configuration. When the differential input voltage $V_1 << V_T$, we can approximate the behavior by linear curve that means $I_{EE}\,tanh\left(\frac{V_1}{2V_T}\right) \approx I_{EE}\left(\frac{V_1}{2V_T}\right)$. Then the value of the collector current difference is, $\Delta I_C = I_{C1} - I_{C2} = I_{EE}\left(\frac{V_1}{2V_T}\right)$. A change that was made to the configuration is depicted in the figure 3.3. A differential amplifier is utilised in the construction of this straightforward modulator circuit. It is possible to employ it as a multiplier so long as V1 is low and significantly lower than 50mV and V2 is higher than VBE (on). The multiplier circuit depicted in the image, however, suffers from a number of drawbacks. The first restriction is that V2 is shifted by VBE when it is turned on. The second issue is that V2 must be positive at all times, which means that the multiplier can only work in two of its four quadrants. The third limitation is that, the tanh (X) is approximately as X, where $X = V_1 / 2V_T$. The first two limitations are overcome in the Gilbert cell.

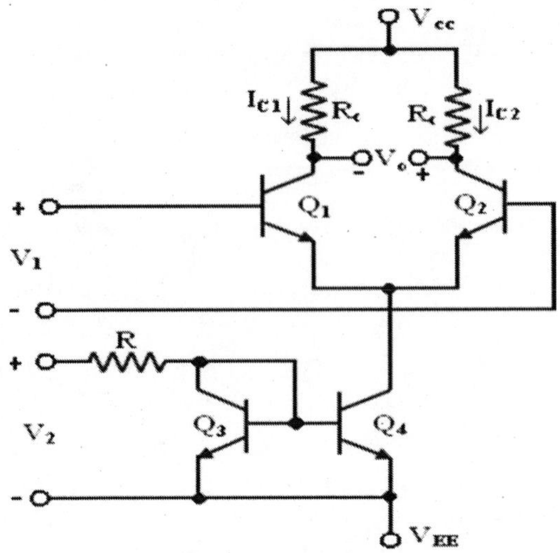

Fig.3.3. Modulation using a differential amplifier

3.3 Gilbert Multiplier cell:

The Gilbert multiplier cell is obtained by making changes to the emitter-coupled cell, which makes it possible to multiply in all four quadrants. Because of this, most balanced multipliers in integrated circuits are based on it. The Gilbert multiplier cell is made up of two emitter-coupled pairs that are cross-coupled to each other and to another emitter-coupled pair.

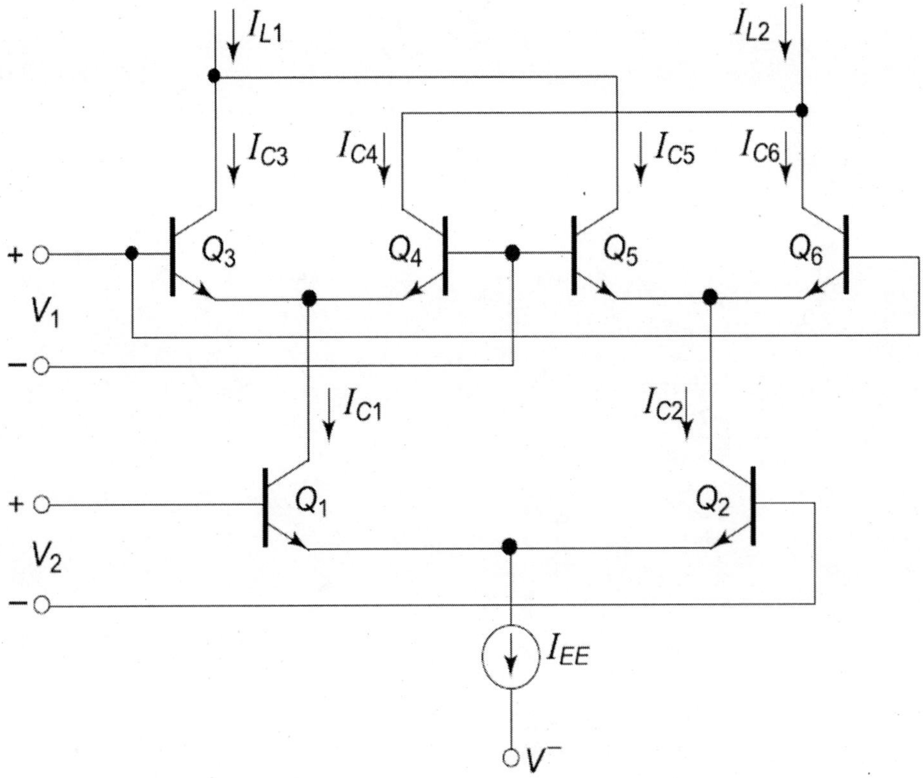

Fig.3.4 Gilbert Multiplier Cell

The differential output current is given by the following equation,

$$\Delta I = I_{L1} - I_{L2} = (I_{C3} - I_{C6}) - (I_{C4} - I_{C5})$$

Simplifying it can be shown that,

$$\Delta I = I_{EE} tanh\left(\frac{V_1}{2V_T}\right) tanh\left(\frac{V_2}{2V_T}\right) \approx I_{EE}\left(\frac{V_1}{2V_T}\right)\left(\frac{V_2}{2V_T}\right)$$

The above equation shows that when V_1 and V_2 are small, the Gilbert Cell in Figure3.4 can be used as a four-quadrant analogue multiplier with the help of current-to-voltage converters.

3.4 Variable Trans conductance Technique

The variable trans conductance method takes advantage of the fact that the trans conductance parameter of a transistor depends on the emitter current bias. You can think of this as a property of the transistor. In figure 3.5, the idea is shown in the form of a simple differential circuit setup.

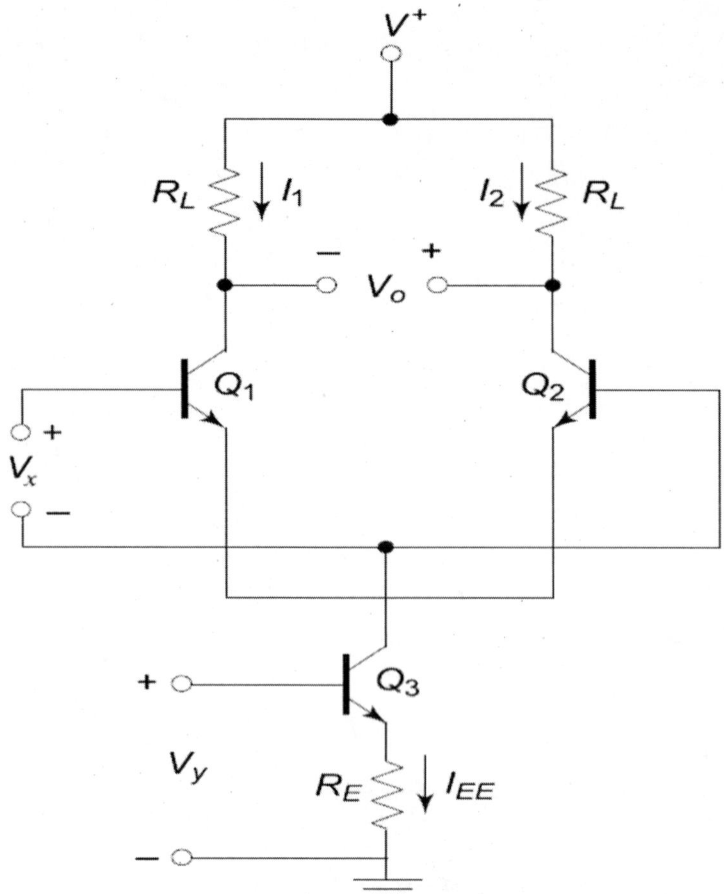

Fig.3.5.Variable Trans conductance Multiplier

From the circuit, $V_0 = g_m R_L V_x$, where $g_m = \frac{I_{EE}}{V_T}$ is the transconductane of the stage. From the bottom Transistor Q3, $V_y = R_E I_{EE}$. Making use of these equations, it can be shown that, $V_0 = \frac{V_y V_x R_L}{V_T R_E}$

Analog Multiplier ICs

An analogue multiplier is a circuit in which the instantaneous value of the product of two separate input voltages determines the proportionate value of the output voltage at any given instant. The most essential uses for these multipliers are modulation and demodulation, multiplication, division, squaring and square-rooting of signals, and multiplication and division of numbers. Integrated circuits featuring op-amps and various other circuit components can serve as a suitable implementation for these analogue multipliers. Figure 3.6 depicts the schematic of a conventional analogue multiplier in the form of an AD633 circuit.

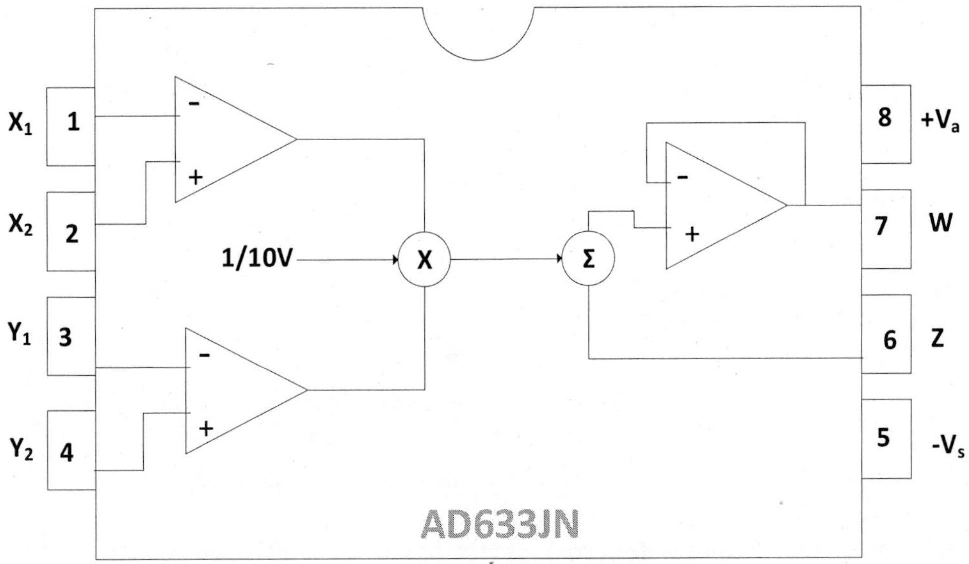

Fig.3.6. Analog Multiplier IC

The AD633 multiplier is an analogue multiplier that utilises all four quadrants. Due to the fact that it has a high input impedance, the effect of loading on the signal source is rendered essentially insignificant by this property. It is able to function with supply voltages that range from -18V to 18V. The IC does not require external components. The calibration performed by the user is not required. The two input signals typically have a range of ten volts or less.The multiplier ICs are used for the following purposes such as Voltage Squarer, Frequency doublers, Voltage divider , Square rooter, Phase angle detector and Rectifier Circuit.

3.5 Phase Locked Loops

The phase-locked loop principle has been applied in a wide variety of contexts, such as stereo decoders for FM (frequency modulation), controls for motor speed, tracking filters, frequency-synthesised transmitters and receivers, FM demodulators, frequency shift keying (FSK) decoders, and the generation of local oscillator frequencies in TV and FM tuners. Other applications include controls for motor speed, tracking filters, and frequency-synthesised transmitters and receivers.

The phase-locked loop is now even available in the form of a single package for easier accessibility. The Signetics SE/NE 560 series, which includes the 560, 561, 562, 564, 565, and 567, are all instances of packages that fall within this category and serve as typical examples. On the other hand, discrete integrated circuits can be used instead of general-purpose integrated circuits in order to construct a phase-locked loop which results in an operation that is more efficient financially.

Figure 3.7 shows the phase-locked loop (PLL) in its basic form. As illustrated in this figure, the phase-locked loop consists of (1) a phase detector, (2) a low-pass filter, and, (3) a Voltage Controlled Oscillator

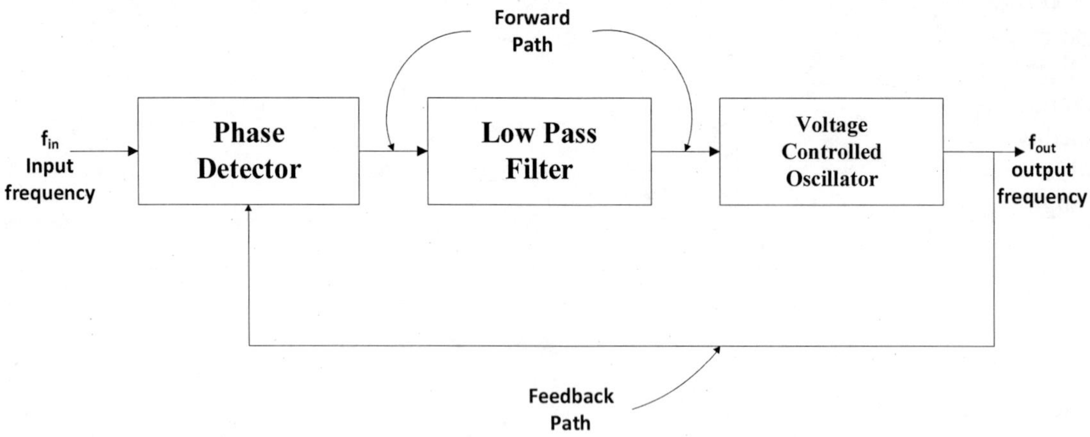

Fig3.7 Block Diagram of Phase Locked Loop

The phase detectors or comparators compare the frequency of the input signal, which is denoted by f_{IN}, to the frequency of the signal that is output, which is marked by f_{OUT}. Due to the fact that the phase detector has an output that is a dc voltage, this number is often referred to as the error voltage due to the characteristics of the signal that it creates. The output of the phase is then routed via a low-pass filter, which gets rid of the high-frequency noise and generates a dc level when it has done so.

This dc level, in turn, is what the voltage-controlled oscillator, often known as the VCO for short, takes as its input. In addition to this, the filter is involved in the process of determining the dynamic properties of the PLL circuit, which contributes to the process. The quantity of direct current (dc) that is supplied to the voltage-controlled oscillator (VCO) has a direct influence on the frequency that is produced by the oscillator. The frequency of the VCO is compared to the frequencies of the input, and the process of modification is repeated until the frequencies of the VCO and the input are identical. In a nutshell, the phase-locked loop iterates between three stages that can be described as free-running, capture, and phase lock. These stages are repeated in a cycle. The phase-locked loop is said to be in the free-running condition when the input is not being applied to it at this point in time. As soon as the input frequency has been adjusted, the frequency of the VCO will start to shift, which is a clear sign that the phase-locked loop is now functioning in the capture mode. When the frequency of the VCO reaches the same level as the input frequency, the phase-locked loop will be in the phase-locked state. The frequency of the VCO will continue to change until it reaches this level. When phase locked, the loop's repeating motion enables it to monitor any change in the input frequency. This is made possible by the phase lock. A discrete phase-locked loop consists of three components: a phase detector, a low-pass filter, and a voltage-controlled oscillator. Each of these components plays a specific function in the loop. The discrete phase-locked loop is the first type of phase-locked loop that we are going to investigate before moving on to the more complex integrated phase-locked-loop circuit.

3.5.1Phase detector:

The input frequency and the frequency of the VCO are both compared by the phase detector. Following this comparison, the phase detector generates a dc voltage whose amplitude is commensurate to the phase difference that exists between the two frequencies. Depending on whether an analogue or digital phase detector was utilised in the phase-locked loop (PLL), it can either be referred to as an analogue or digital variant. The PLL stands for phase-locked loop. Even though analogue phase detectors are used in the most majority of monolithic PLL integrated circuits, digital phase detectors make up the large majority of discrete phase detectors that are currently in use. This is because digital phase detectors are more accurate than their analogue counterparts. This is partly because digital phase detectors are significantly simpler to operate than their analogue counterparts.

An analogue phase detector that is tried and true is typically represented by a double-balanced mixer. On the other hand, some illustrations of digital phase detectors include the following:

1. Exclusive-OR phase detector
2. Edge-triggered phase detector
3. Monolithic phase detector (such as type 4044)

The following fig shows Exclusive-OR phase detector:

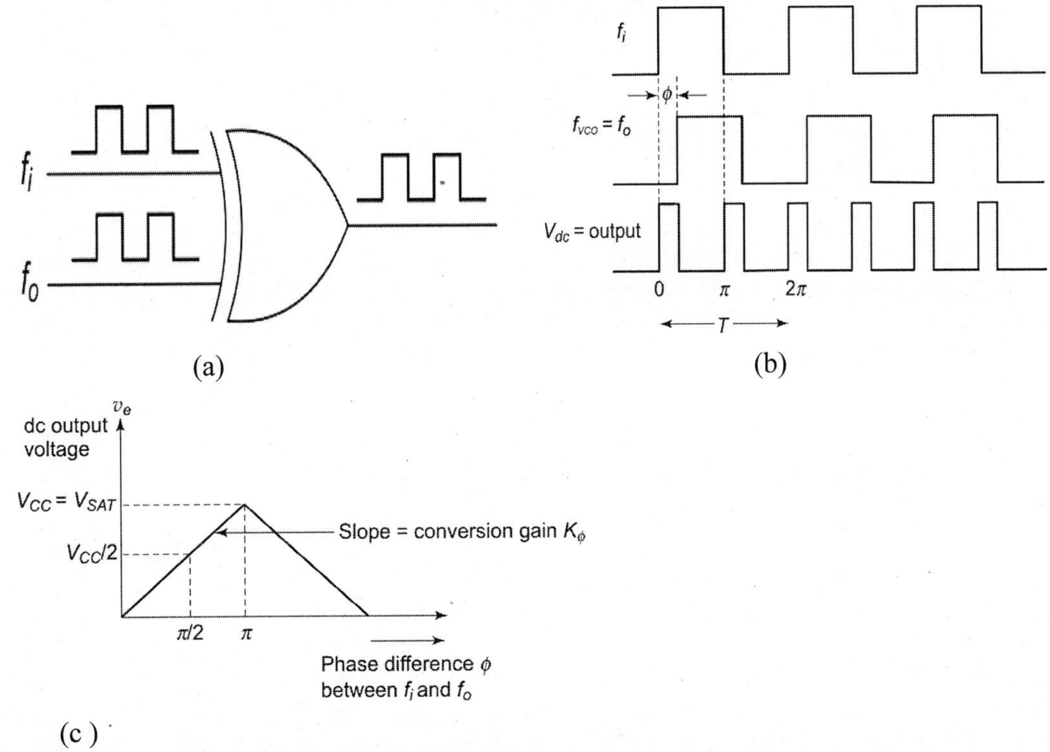

Fig.3.8 (a) Exclusive-OR phase detector: connection and logic diagram. (b) Input and output waveforms. (c) Average output voltage versus phase difference between fIN and fOUT curve.

3.5.2 Low-pass filter

At the output of the phase detector, there will be high-frequency components that need to be filtered out. The task of the low-pass filter is to filter out the noise in addition to the high-frequency components.

Even more critically, the low-pass filter is in charge of managing the dynamic features of the phase-locked loop. This is an extremely important function. Some examples of these characteristics include the capture and lock ranges, the bandwidth, and the transient responsiveness. The PLL system monitors a frequency range known as the lock range for any changes that may occur in the input frequency known as f_{IN}. It is the range of frequencies over which these shifts occur, and its definition is that range. Lock range and tracking range are both terms that can be used interchangeably in this context. The frequency range across which the PLL is capable of capturing phase lock is referred to as the capture range. In other words, the capture range is the frequency range over which the PLL can achieve phase lock. It really shouldn't come as a big surprise that the lock range is going to be significantly greater than the capture range.

3.5.3 Voltage-controlled oscillator:

The phase-locked loop contains three components, the third of which is the voltage-controlled oscillator. A direct proportional relationship exists between the amount of voltage that is introduced into the VCO and the frequency that is generated by the VCO. Because it creates simultaneous square wave and triangle wave outputs as a function of the input voltage, the Signetics NE/SE 566 VCO is a typical example of a voltage-controlled oscillator (VCO). It is manufactured by the company Signetics. Figure displays the functional block diagram of the VCO for your viewing pleasure. The frequency of the oscillations is determined by the three resistors R_1 that are located on the outside of the circuit, the capacitor C_1, and the voltage V_C that is connected to control terminal 5.

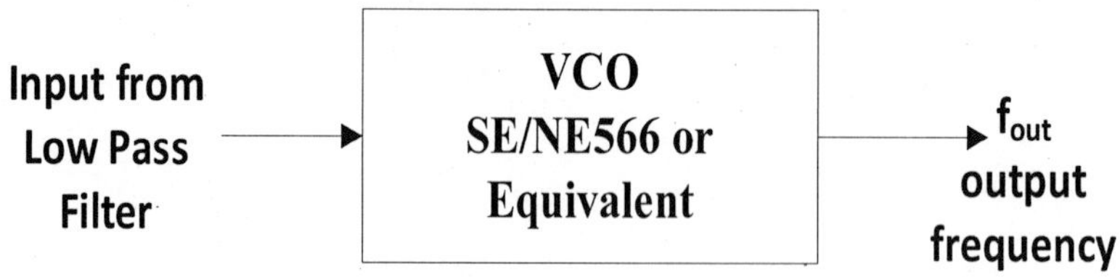

Fig.3.9 VCO Block Diagram

3.6 Monolithic Phase Lock Loops IC 565:

Monolithic PLLs are introduced by signetics as SE/NE 560 series and by national semiconductors LM 560 series. The 560/561/562 series was the first to be created as a monolithic PLL with any practical application. The LM560 series produced by National Semiconductor Corp. and the SE/NE560 series produced by Signetics are two examples of major monolithic PLLs series. The SE/NE560, 561, 562, 564, 565, and 567 series are distinguished from one another primarily by the operating frequency ranges, the minimum power supply need, and the maximum bandwidth adjustment ranges that they offer. In this section, the LM565 phase-locked loop (PLL), which is the most widely used PLL, is discussed. These PLL circuits include a temperature-compensated VCO, a phase detector, and a provision for connecting an external RC circuit to carry out the loop-filtering function.

Figure 3.10 and 3.11 show, respectively, the pin diagram and the block diagram of the IC 565. Both a 14-pin DIP package and a 10-pin Metal Can container are accessible for use with the IC 565. In accordance with what is depicted in Fig. 10.15(b), the output of the phase detector is sent into a differential amplifier. Internally to the VCO is where a single terminated output that has been dropped through R is attached. Additionally, you can access it as a demodulated output on pin 7 of the PLL. Resistor R is an integral component of the low-pass filter. The low-pass filter is formed by the capacitor C, which is located between pins 7 and 10, and the resistor R, which has a rating of 3.6 kW. Capacitor C must to be of sufficient size so as to avoid fluctuations in the voltage that is output by the demodulator at pin 7.In configurations in which twin power supplies are utilised, the power supply need for the 565 might be any value within the range of 6 V to 12 V. When only one power supply is being used, the voltage of the supply might range anywhere from +12 V to +24 V.

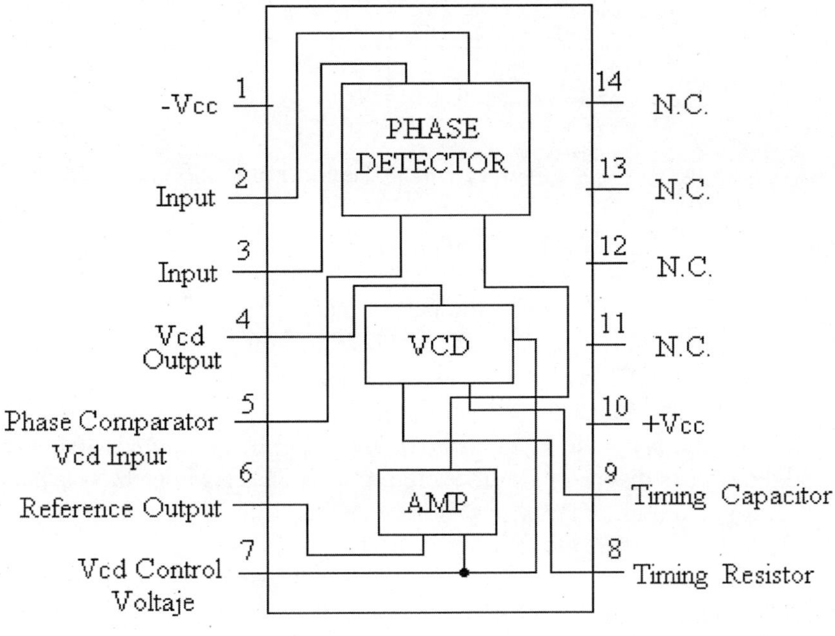

Fig3.10 Pin configuration of IC 565

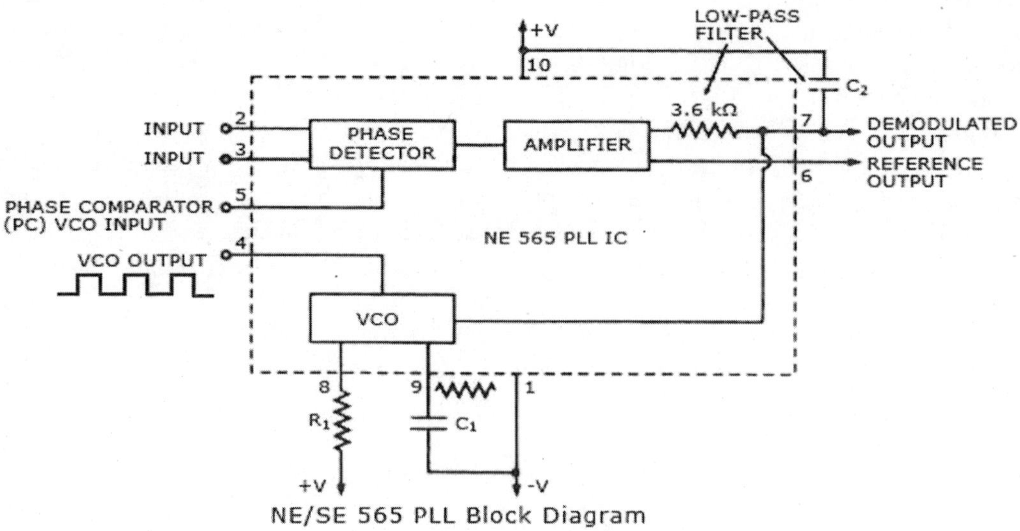

Fig3.11 Block Diagram of IC 565

3.7 Application of PLL AM demodulation

Demodulating AM signals can be accomplished with the help of a PLL, as demonstrated in the following figure 3.12. The PLL has been tuned to match the carrier frequency of the AM signal that is coming in. The multiplier receives the unmodulated output of the VCO, which has the same frequency as the carrier signal. Due to the fact that the output of the VCO is always 900 before it is fed into the multiplier. Because of this, the demodulated output is created by filtering high frequency components using the LPF. This applies to both the signals that are applied to the multiplier and the signals that represent the difference. A PLL AM detector displays a high degree of selectivity and noise immunity, which is not feasible with traditional peak detector type AM modulators due to the fact that the PLL only reacts to the carrier frequencies that are extremely close to the VCO output. This is because the PLL only responds to the frequencies that are very close to the VCO output.

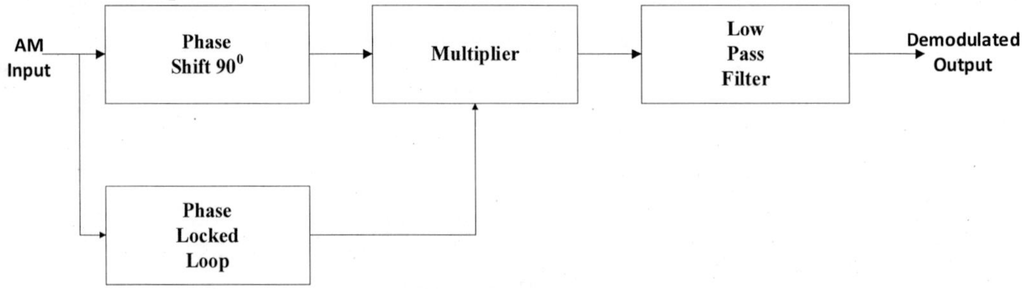

Figure 3.12 AM Demodulation using PLL

3.8 FM Demodulation

The VCO will track the instantaneous frequency of the input signal if the PLL is locked to an FM signal. The filtered error voltage is the demodulated FM output. This voltage controls the VCO and ensures that it remains locked with the input signal. The linearity of the demodulated output can be

directly attributed to the transfer characteristics of the VCO. It is possible to realise very linear FM demodulators due to the highly linear nature of the VCO that is employed in IC PLL.
Figure 3.13 presents a block schematic of the process of FM Demodulation.

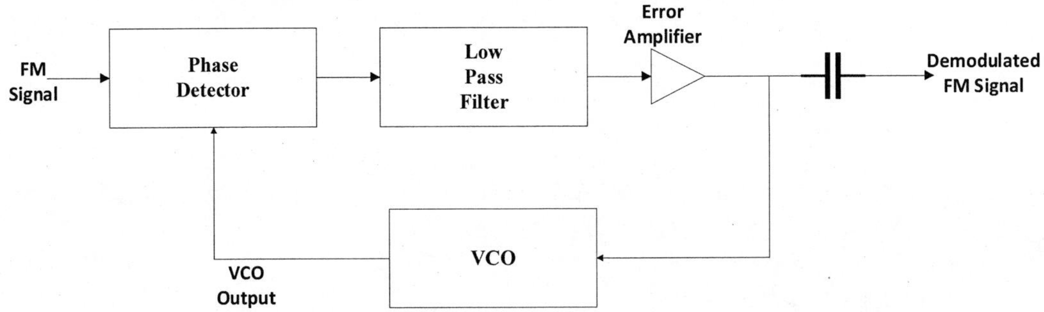

Fig.3.13.FM demodulation process.

3.9 FSK modulation and demodulation

A carrier frequency is utilised in digital data communication in order to facilitate the transmission of binary data. In order to distinguish between the logic 1 and logic 0 states of a binary data signal, it makes use of two distinct carrier frequencies. The acronym FSK stands for frequency shift keying, which describes this method of data transfer. In order to recover the original binary data that was sent during this data transmission, the receiving end must first transform two carrier frequencies into the values 1 and 0. Demodulation of frequency shift keying is the name given to this particular method. FSK block diagram a phase-locked loop (PLL) can function as a frequency shift keying demodulator, as demonstrated in Figure 3.14 of the design.

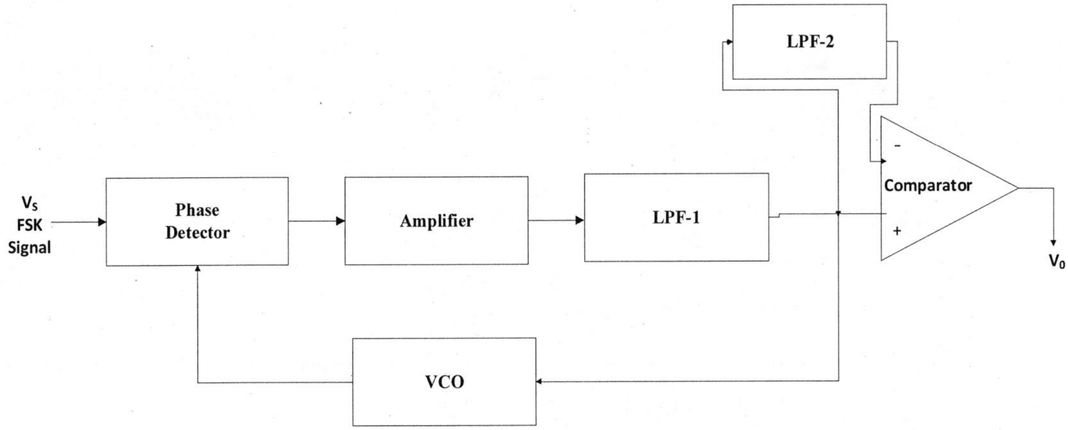

Fig.3.14. FSK Modulation and demodulation

It is quite similar to the PLL demodulator that is used for analogue FM signals; the only difference is that it also includes a comparator that generates a reconstructed digital output signal. Consider for a moment that there are two frequencies, and that one of those frequencies (f1) is denoted by the numeral "0," while the second frequency (f2) is denoted by the numeral "1." In the event that the PLL continues to be locked into the FSK signal at both f1 and f2; the control voltage for the VCO, which is also supplied to the comparator, will be written out as

$V_{C1} = (f_1 - f_0) / K_v$ and

$V_{C2} = (f_2 - f_0) / K_v$, respectively.

where K_v is the voltage to frequency transfer coefficient of the VCO. The difference between the two control voltage levels will be

$\Delta VC = (f2 - f1) / Kv$.

The comparator's reference voltage is produced from the additional low pass filter, and its value is modified so that it is halfway between V_{C1} and V_{C2} in the voltage range.As a result, the output of the comparator is written as '0' for VC1 and '1' for V_{C2}.

3.10 Frequency Synthesizer

The phase-locked loop (PLL) can serve as the foundation for the construction of a frequency synthesiser that is able to generate an accurate sequence of frequencies that are obtained from a crystal-controlled oscillator that is steady.

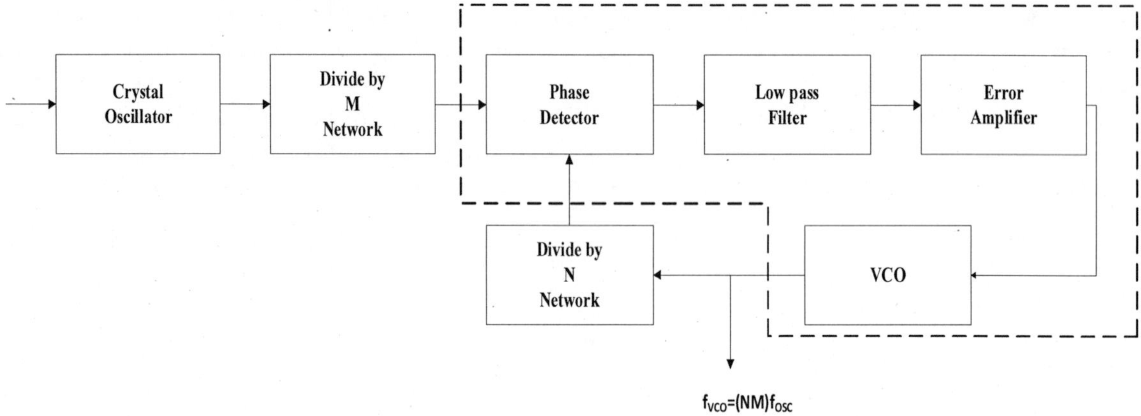

Fig.3.15.Frequency Synthesizer

The block diagram of the frequency synthesiser is displayed here in Figure 3.15. The only difference between it and the frequency multiplier circuit is the addition of the divided-by-M network at the phase lock loop's input. f_{osc} is the frequency of the crystal-controlled oscillator, and M is an integer factor. The divider network takes the frequency of the crystal-controlled oscillator and divides it by M to generate the frequency f_{osc}/M. The frequency of the VCO, denoted by f_{VCO}, is similarly divided by the factor N used by the divider network to produce the resultant frequency, which is equal to f_{vco} divided by N. When the PLL has stabilised itself on the frequency of the divided-down oscillator, we will obtain $f_{osc}/M = f_{vco}/N$; hence, f_{vco} will equal $(N/M)f_{osc}$.A wide variety of different frequencies can be generated, all of which are derived from the crystal-controlled oscillator, simply by setting the divider counts to the desired values.

UNIT SUMMARY

- An emitter coupled pair can be used as a multiplier with the change in collector current
- The Gilbert multiplier cell which is a modification of the emitter coupled cell forms the basis of many IC multipliers and this cell allows four-quadrant multiplication
- Variable transconductance multiplier makes use of the dependence characteristic of the transistor transconductance parameter on the emitter current bias applied.
- The phase locked loop, often known as PLL, is a closed loop feedback system that is an essential component of a linear system. The output frequency and phase of the PLL are synchronised with the frequency and phase of the signal that is being input into the system.
- It has the ability to detect the phases of the two signals and to attenuate the difference between them when there is such a difference. The phase detector/comparator, the lowpass filter, the error amplifier (A), and the voltage controlled oscillator (VCO) are the primary building parts of the phase-locked loop.
- When the PLL is in its unlocked state, the VCO functions at a frequency referred to as centre frequency or free running frequency. This frequency corresponds to an applied voltage of 0 V dc at its control input. When the PLL is locked, it operates at a frequency of fc.
- The capture range of a phase-locked loop (PLL) is defined as the range of input frequencies around the centre frequency that fall within the parameters of the loop's ability to lock up after being in an unlocked state. The total amount of time necessary for the loop to get aligned with the input signal is referred to as the pull-in time.
- When the lock is suddenly lost because of a significant interfering transient signal, the lowpass loop-filter works as a memory for the loop and filters out the difference frequency components that come from interference signals that are far removed from the centre frequency.
- When the filter's bandwidth is narrowed, its ability to reject out-of-band signals is improved. However, this also has the effect of reducing the capture range, which causes the pull-in time to lengthen and causes the loop phase margin to narrow.
- The PLL 565 is used in applications such as frequency multiplication, frequency division, AM Detection,FM Detection, FSK modulation and demodulation, and frequency synthesising.

Multiple Choice Quesztions

1. PLL uses the following logic gate as PD

   (a) OR gate (b) NAND gate (c) EX-NOR gate (d) XOR gate

   [Ans. (d)]

2. Phase-locked loop uses the following filter

   (a) HPF (b) BPF

   (b) Band rejection filter (d) LPF

   [Ans. (d)]

3. Phase detector is also known as

    (a) Subtractor (b) Adder (c) Divider (d) Multiplier

[Ans. (d)]

4. Output voltage of VCO is proportional to

    (a) Input voltage (b) Input signal frequency

    (b) Input time (d) None of the above

[Ans. (a)]

Short and Long Answer Type Questions

1. Explain the operation of the Gilbert Cell
2. Draw and explain the operation of Multiplier IC.
3. Explain the pin diagram of IC 565
4. Explain any 2 applications of PLL
5. With necessary diagrams, explain the Frequency Synthesizer

REFERENCES AND SUGGESTED READINGS

1. Coughlin, R. F. and Frederick F. Driscoll (2001). Operational Amplifiers and Linear Integrated Circuits, Sixth Edition. Upper Saddle River, NJ: Prentice Hall.
2. Dailey, Denton J. (1989) Operational Amplifiers and Linear Integrated Circuits: Theory andApplications. New York: The McGraw-Hill Companies.
3. Franco, Sergio (2002). Design with Operational Amplifiers and Analog Integrated Circuits.New York: The McGraw-Hill Companies.
4. Gayakwad, Ramakant A. (1999). Op-Amps and Linear Integrated Circuits. Fourth edition.Upper Saddle River, NJ: Prentice Hall Inc.
5. Jacob, J. Michael (1996) Applications and Design with Analog Integrated Circuits. Upper
6. Saddle River, NJ: Prentice Hall Inc.
7. *Ramakanth A. Gayakwad* (2000),Op-Amps and Linear Integrated Circuits -, *Prentice-Hall, 4th edition*.
8. S.Salivahanan,V S Kanchana Bhaskaran(2015),Linear Integrated Circuits, Second Edition, Tata McGrawa Hill Edition.
9. B.Visweswara Rao(2015), Linear Integrated Circuits, First Edition, Pearson India Education Services Pvt. Ltd

Dynamic QR Code for Further Reading

4 Analog to Digital and Digital to Analog Converters

UNIT SPECIFICS

Through this unit we have discussed the following aspects:

- *Different types of data converters;*
- *Different types of Analog to Digital Converters;*
- *Different types of Digital to Analog Converters;*
- *Discussion on specifications of converters;*

RATIONALE

The data converters are responsible for transforming one type of data into another. Real-world processes generate analogue signals, which convey information about process variables such as voltage, current, charge, temperature, and pressure. These signals are produced by real-world processes. It's possible for the flow of such information to be either slow or extremely quick. Using just analogue technology makes it difficult to accurately store, handle, compare, calculate, and retrieve such data. Using analogue technology also makes it difficult to recover such data. However, by utilising digital methods, computers are able to carry out these procedures in a speedy and effective manner. For this reason, it is important to convert the analogue signals that are produced by the various transducers into their corresponding digital data. These digital data then serve as the input for the digital systems. Because of this, there was a growing demand for the conversion of analogue signals into digital data. Because the computers also need to communicate with humans and physical processes through the use of analogue signals, the technique of converting digital data to analogue data was required. The most common D/A and A/D conversion techniques, D/A converter and A/D converter integrated circuits, delta modulators and demodulators, followed by sigma-delta converters, and frequently used ICs for D/A and A/D conversions are covered in this chapter.

PRE-REQUISITES

Electronic devices and circuits, Circuit theory

UNIT OUTCOMES

List of outcomes of this unit is as follows:

U4-O1: Describe conversion mechanism

U4-O2: Explain the different specifications of converters

U4-O3: Explain Analog to Digital Converters

U4-O4: Discuss about Digital to Analog Converters

U4-O5: Comparison of the different converters

Unit-4 Outcomes	**EXPECTED MAPPING WITH COURSE OUTCOMES** *(1-Weak Correlation; 2- Medium Correlation; 3-Strong Correlation)*					
	CO-1	**CO-2**	**CO-3**	**CO-4**	**CO-5**	**CO-6**
U4-01	2	3	3	3	2	3
U4-02	1	3	3	3	1	2
U4-03	2	3	3	3	2	3
U4-04	1	3	3	3	1	2
U4-05	-	3	3	3	1	3

4.1 BASIC DAC TECHNIQUES

The input of the block diagram is binary data i.e, 0 and 1,it contain 'n' number of input bits designated as $d_1,d_2,d_3,.....d_n$ this input is combined with the reference voltage called V_{dd} to give an analog output. Where d_1 is the MSB bit and d_n is the LSB bit

$$V_0 = V_{dd}(d_1 * 2^{-1} + d_2 * 2^{-2} + d_3 * 2^{-3} + ... d_n * 2^{-n})$$

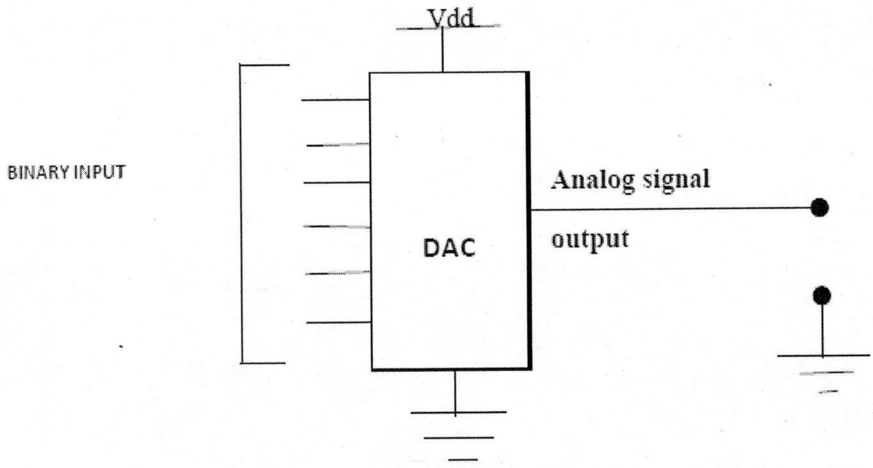

Fig.4.1: Basic DAC diagram

4.2 Weighted Resistor:

Fig.4.2 shows a simplest circuit of weighted resistor. It uses a summing inverting amplifier. It contains n- electronic switches (i.e. 4 switches) and these switches are controlled by binary input bits d_1, d_2, d_3, d_4. If the binary input bit is 1 then the switch is connected to reference voltage $-V_{REF}$, if the binary input bit is 0 then the switch is connected to ground.

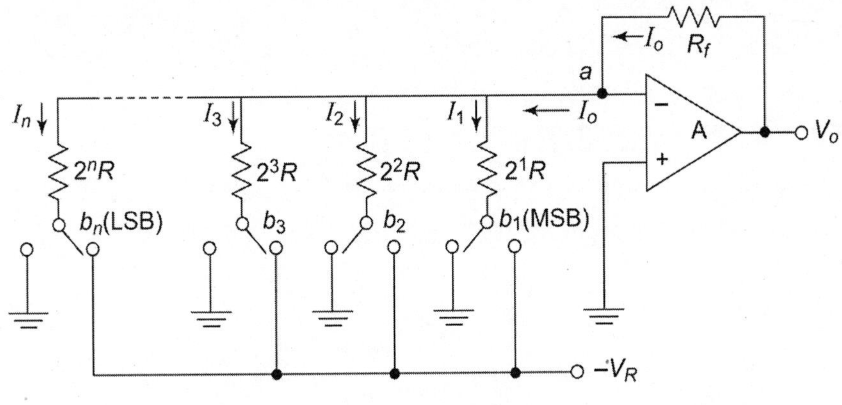

Fig. 4.2:: simple 4-bit weighted resistor

The output current equation is $I_o = I_1 + I_2 + I_3 + I_4$

$$I_0 = V_{REF}(d_1 * 2^{-1} + d_2 * 2^{-2} + d_3 * 2^{-3} + d_4 * 2^{-4})$$

The transfer characteristics are shown below for a 3-bit weighted resistor in Fig.4.3

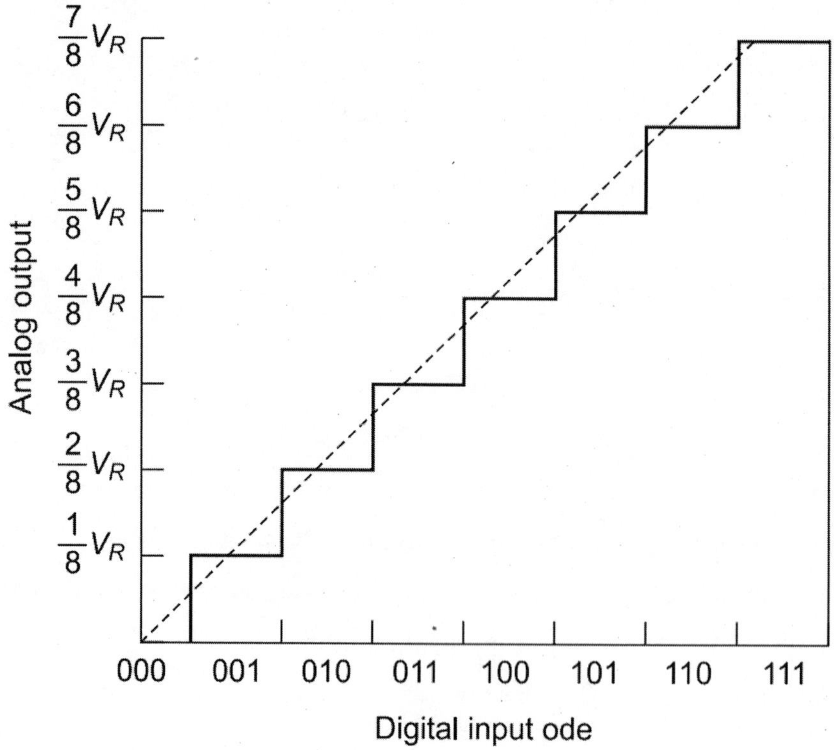

Fig 4.3: Transfer characteristics of 3-bit weighted resistor

4.2.1 Disadvantages of Weighted resistor D/A converter:

Wide range of resistor's are required in this circuit and it is very difficult to fabricate such a wide range of resistance values in monolithic IC. This difficulty can be eliminated using R-2R ladder network.

4.3 R-2R LADDER DAC

Wide range of resistors required in binary weighted resistor type DAC. This can be avoided by using R-2R ladder type DAC. The circuit of R-2R ladder network is shown in fig.4.4.

The basic theory of the R-2R ladder network is that current flowing through any input resistor (2R) encounters two possible paths at the far end. The effective resistances of both paths are the same (also 2R), so the incoming current splits equally along both paths. The half-current that flows back towards lower orders of magnitude does not reach the op amp, and therefore has no effect on the output voltage. The half that takes the path towards the op amp along the ladder can affect the output. The inverting input of the op-amp is at virtual earth. Current flowing in the elements of the ladder network is therefore unaffected by switch positions.

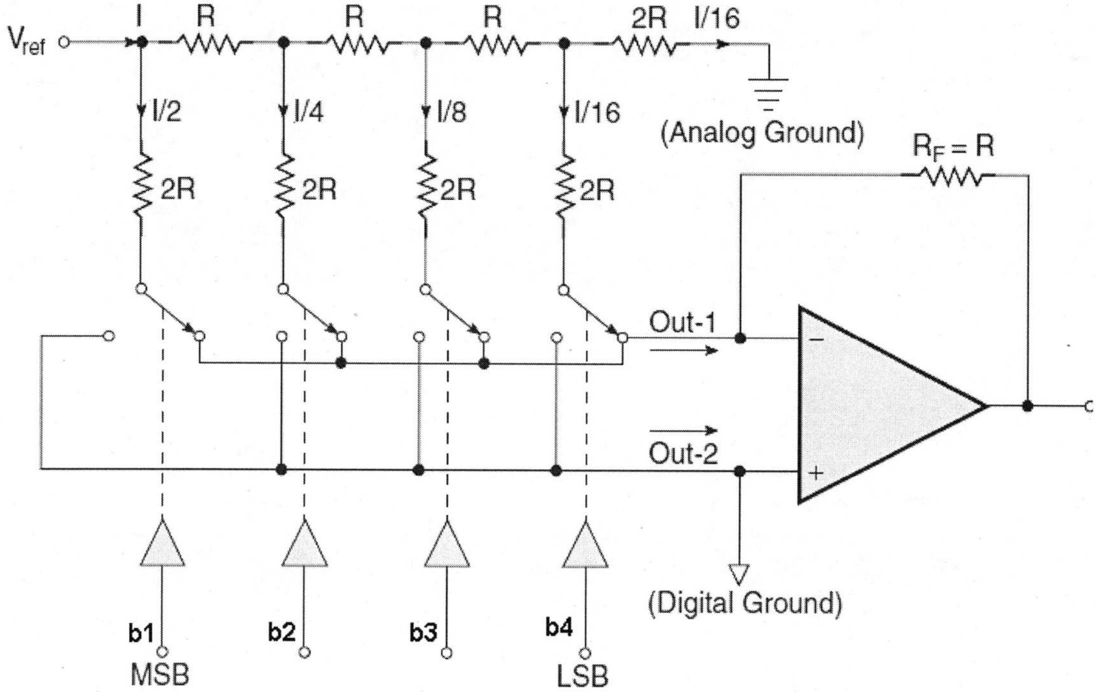

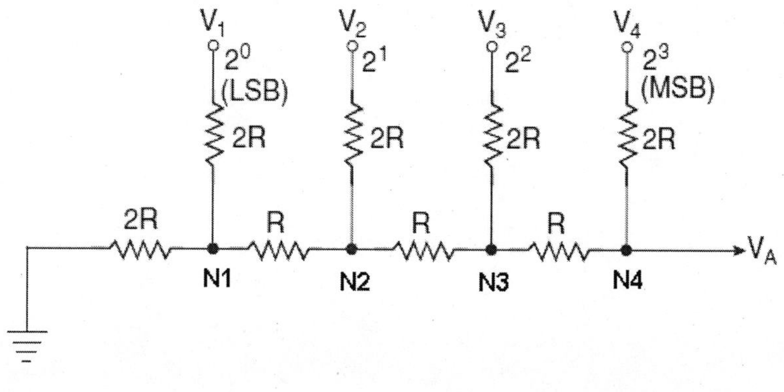

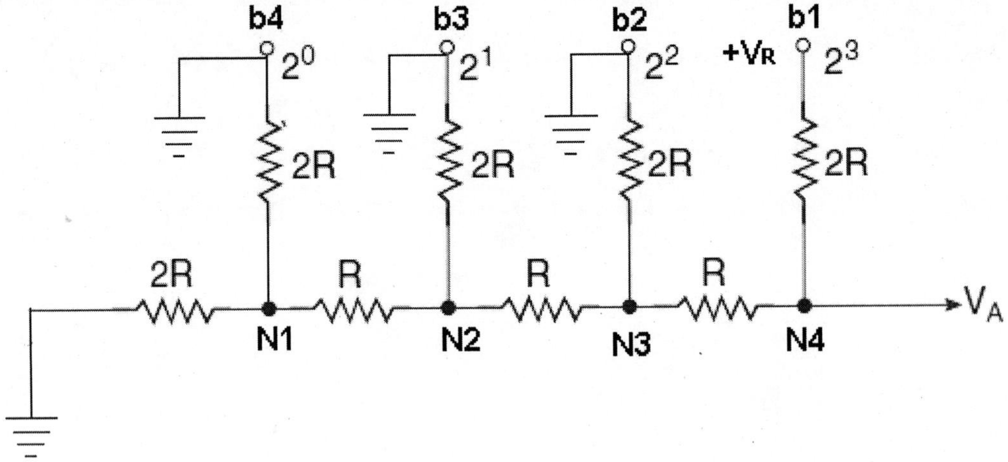
b4
b3
b2
b1
2^0
2^1
2^2
$+V_R$
2^3
2R
2R
2R
2R
2R
R
R
R
N1
N2
N3
N4
V_A

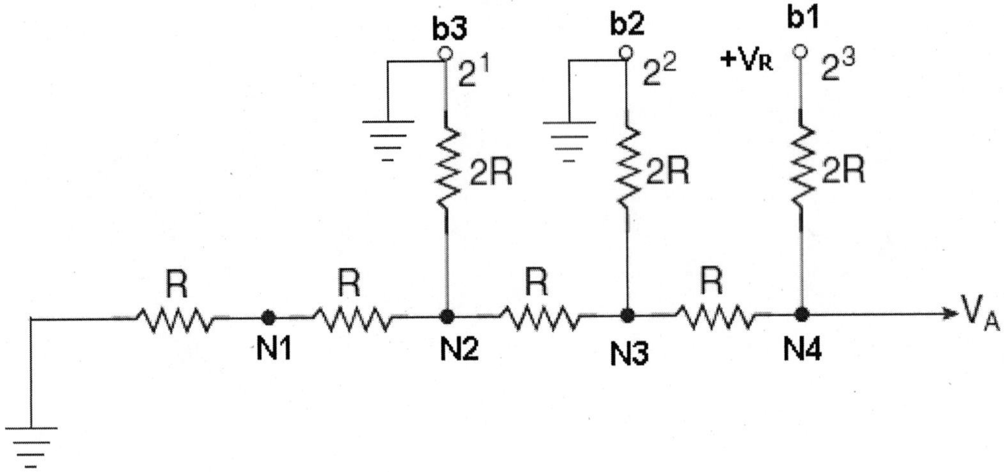
b3
b2
b1
2^1
2^2
$+V_R$
2^3
2R
2R
2R
R
R
R
R
N1
N2
N3
N4
V_A

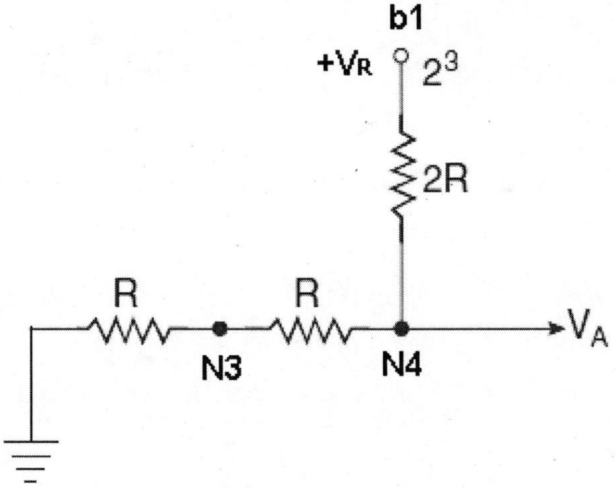

Fig 4.4: A 4-bit R-2R Ladder DAC

If we label the bits (or inputs) bit 1 to bit N the output voltage caused by connecting a particular bit to V_r with all other bits grounded is:

$$V_{out} = \frac{V_r}{2N}$$

where N is the bit number. For bit 1, $V_{out} = V_r/2$, for bit 2, $V_{out} = V_r/4$ etc.Since an R/2R ladder is a linear circuit, we can apply the principle of superposition to calculate V_{out}. The expected output voltage is calculated by summing the effect of all bits connected to Vr. For example, if bits 1 and 3 are connected to V_r with all other inputs grounded, the output voltage is calculated by:$V_{out} = (V_r/2)+(V_r/8)$ which reduces to $V_{out} = 5V_r/8$.An R/2R ladder of 4 bits would have a full-scale output voltage of 1/2 +1/4 + 1/8 + 1/16 = $15V_r/16$ or 0.9375 volts (if V_r=1 volt) while a 10-bit R/2R ladder would have a full- scale output voltage of 0.99902 (if V_r=1 volt).

4.4 INVERTED R-2R LADDER DAC

In weighted resistor and R-2R ladder DAC the current flowing through the resistor is always changed because of the changing input binary bits 0 and 1. More power dissipation causes heating, which in turn creates non-linearity in DAC. This problem can be avoided by using Inverted r-2r ladder DAC. In this MSB and LSB is interchanged. Here each input binary word connects the corresponding switch either to ground or to the inverting input terminal of op-amp which is also at virtual ground. When the input binary in logic 1 then it is connected to the virtual ground, when input binary is logic 0 then it is connected to the ground i.e. the current flowing through the resistor is constant.

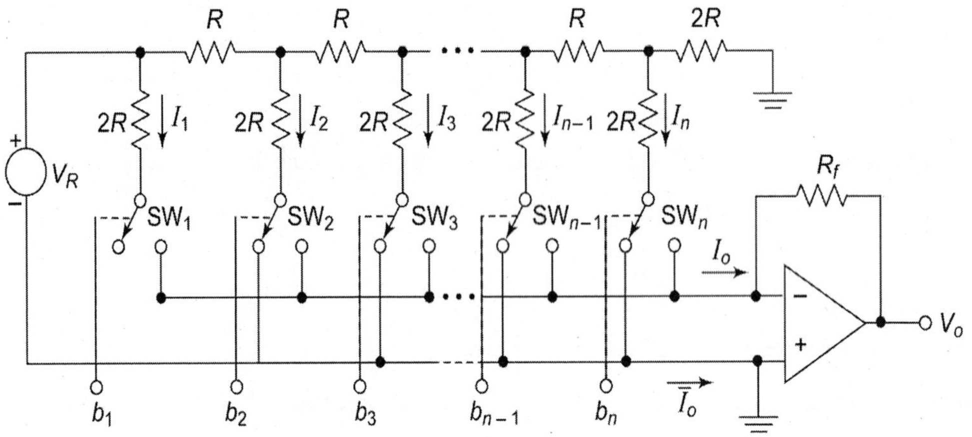

Fig 4.5 : Inverted R-2R ladder

4.5 DIFFERENT TYPES OF ADC'S

It provides the function just opposite to that of a DAC. It accepts an analog input voltage Va and produces an output binary word d1, d2, d3....dn. Where d1 is the most significant bit and dn is the least significant bit.

ADCs are broadly classified into two groups according to their conversion techniques

1) Direct type

2) Integrating type

Direct type ADCs compares a given analog signal with the internally generated equivalent signal. This group includes

i) Flash (Comparator) type converter

ii) Successive approximation type converter

iii) Counter type

iv) Servo or Tracking type

Integrated type ADCs perform conversion in an indirect manner by first changing the analog input signal to linear function of time or frequency and then to a digital code.

4.6 FLASH (COMPARATOR) TYPE CONVERTER:

A flash ADC or direct-conversion ADC has a bank of comparators that sample the input signal in tandem, each firing for their decoded voltage range. The comparator bank provides input to a logic circuit, which generates a code for each voltage range. Direct conversion is extremely rapid, capable of gigahertz sampling rates, but typically has only 8 bits of resolution or less, because the number of comparators required, 2N - 1, doubles with each new bit, necessitating a huge, expensive circuit. This type of ADC has a large die size, a high input capacitance, a high power dissipation, and is prone to producing output glitches (by generating an out-of-sequence code). Scaling to newer sub-micrometre technologies is ineffective because device mismatch is the primary design constraint. They are frequently employed in optical storage for video, wide band communications, or other rapid signals.

A Flash ADC (also known as a direct conversion ADC) is a form of analog-to-digital converter that compares the input voltage to successive reference voltages using a linear voltage ladder with a comparator at each "rung" of the ladder. These reference ladders are frequently made up of several resistors; however, newer implementations demonstrate that capacitive voltage division is also achievable. The output of these comparators is typically supplied into a digital encoder, which turns the inputs into a binary value (the sum of the comparator outputs can be thought of as a unary value).

This circuit, often known as the parallel A/D converter, is the simplest to comprehend. It is made up of a sequence of comparators, each of which compares the input signal to a different reference voltage. The comparator outputs are connected to the inputs of a priority encoder circuit, which results in a binary output.

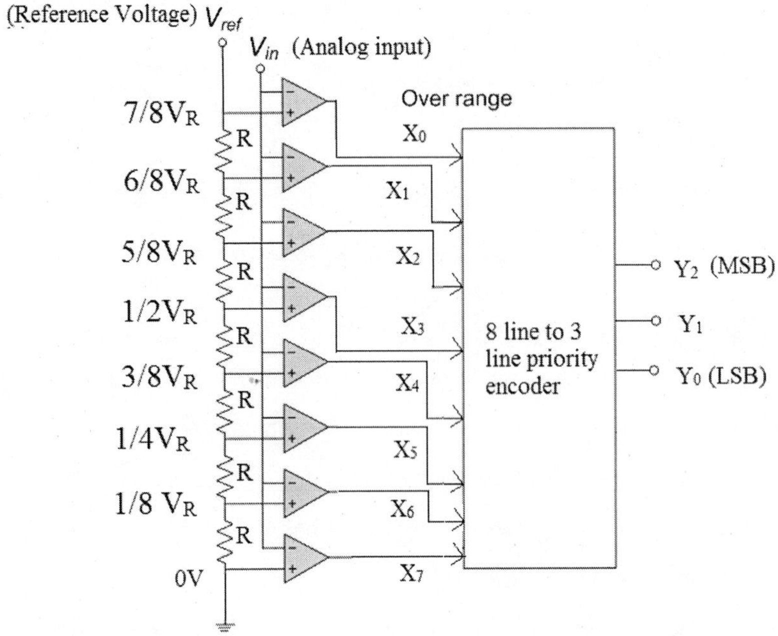

Fig.4.6 : Flash (parallel comparator) type ADC

V_R is a stable reference voltage provided by a precision voltage regulator, which is not represented in the schematic. As the analogue input voltage at each comparator surpasses the reference voltage, the comparator outputs will sequentially saturate to a high state. The priority encoder produces a binary number based on the highest-order active input while disregarding all other active inputs.

4.7 Counter type A/D converter

The counter in the figure4.7 is reset to zero count by a reset pulse. The binary counter counts the clock pulses when the reset pulse is released. These pulses are routed through the AND gate, which is activated by the high voltage comparator output. The number of pulses counted grows over time.

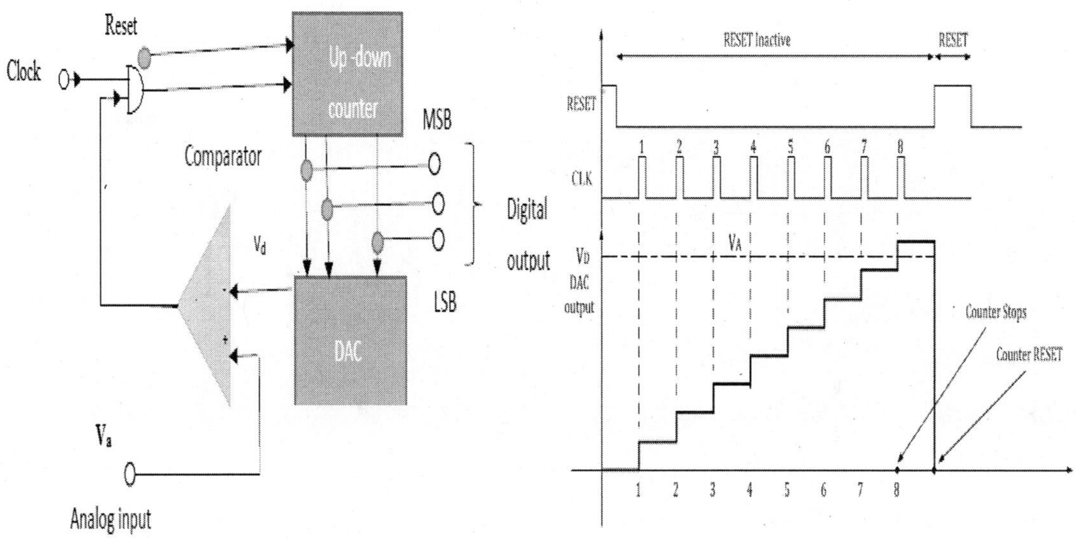

Fig.4.7 : Countertype A/D converter

The binary word representing this count is fed through a D/A converter, the output of which is a stair case. The comparator compares the analogue output Vd of the DAC to the analogue input input Va. If Va>Vd, the comparator output becomes high, and the AND gate is enabled, allowing clock pulses to be transmitted to the counter. When Va Vd, the comparator output goes low, and the AND gate is disabled. This brings the counting to a halt, allowing us to obtain the digital data.

4.8 Servo Tracking A/D Converter :

An improved version of counting ADC is the tracking or servo converter shown in fig4.8. The circuit consists of an up/down counter with the comparator controlling the direction of the count.

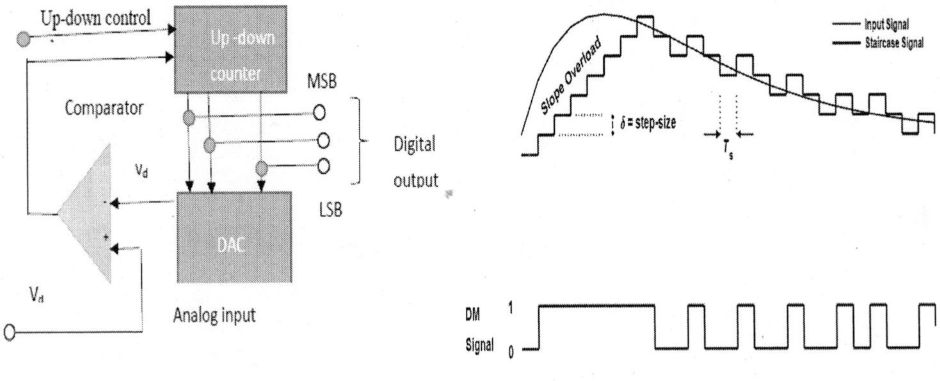

Fig:4.8 : (a) A tracking A/D converter (b) waveform associated with a tracking A/D converter

The DAC's analogue output, V_d, is compared to the analogue input, Va. If the input V_a is greater than the DAC output signal, the comparator output becomes high, and the counter begins to count up. With each incoming clock pulse, the DAC output grows, and when it exceeds V_a, the counter flips direction and counts down.

4.9 Successive-Approximation ADC:

The so-called successive-approximation ADC is one technique of overcoming the drawbacks of the digital ramp ADC. The sole difference between this design and the one depicted in the figure4.9 is the use of a highly specific counter circuit known as a successive-approximation register.

Instead of counting up in binary order, this register counts by attempting all bit values beginning with the most significant bit and ending with the least significant bit. The register continuously examines the comparator's output to detect if the binary count is less than or larger than the analogue signal input, and adjusts the bit values accordingly. The register counts in the same way as the "trial-and-fit" method of decimal-to-binary conversion, in which different bit values are attempted from MSB to LSB to produce a binary number that equals the original decimal number. The advantage of this counting approach is that it produces much faster results: the

DAC output converges on the analogue signal input in much greater increments than a standard counter's 0-to-full count sequence.

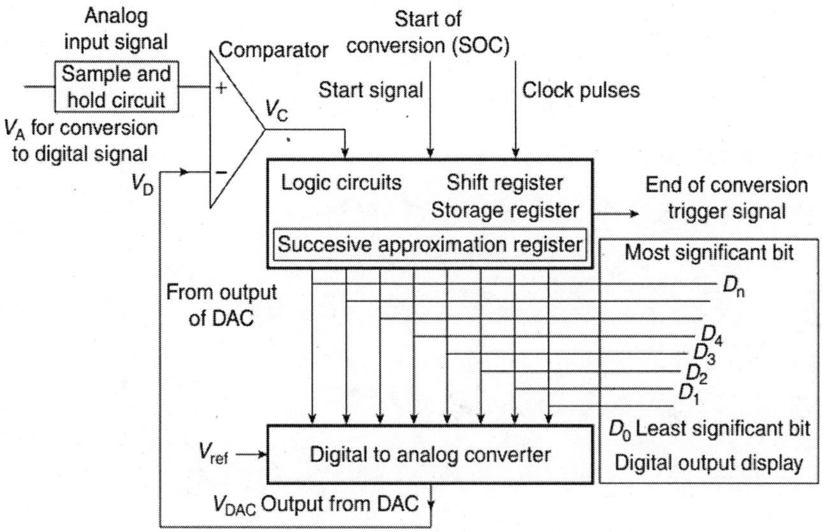

Fig:4.9 : Successive approximation ADC circuits

The successive approximation analog to digital converter circuit typically consists of four chief subcircuits

1. A sample and hold circuit is used to acquire the input voltage, which is denoted by Vin.
2. An analogue voltage comparator that compares Vin to the output of the internal DAC and sends the result of that comparison to the successive approximation register (also known as SAR).
3. A consecutive approximation register sub circuit that has been built to supply a digital code that is approximately equivalent to Vin to the internal DAC.
4. A digital-to-analog converter (DAC) that is located internally and supplies the comparator with an analogue voltage that is equivalent to the digital code that is output by the SAR so that it can be compared with Vin.

When the successive approximation register is first set up, the most significant bit (MSB) is set to the value of a digital 1, as this is the convention. This code is then sent into the DAC, which then sends the analogue equivalent of this digital code (Vref/2) into the comparator circuit so that it may be compared with the sampled input voltage. The comparator will force the SAR to reset this bit if the current analogue voltage is greater than Vin; if it is not, the bit will remain set to 1. After that, the next bit is made to have a value of one, and the previous test is repeated. This binary search is carried on until each bit in the SAR has been examined. The code that is generated as a result is the digital approximation of the sampled input voltage, and it is finally produced by the DAC at the conclusion of the conversion process (EOC).Mathematically, let $V_{in} = xV_{ref}$, so x in [-1, 1] is the normalized input voltage. The objective is to approximately digitize x to an accuracy of $1/2^n$. The algorithm proceeds as follows:

1. Initial approximation $x_0 = 0$.
2. ith approximation $x_i = x_{i-1} - s(x_{i-1} - x)/2^i$. where, s(x) is the signum-function(sgn(x)) (+1 for $x \geq 0$, -1 for $x < 0$). It follows using mathematical induction that $|xn - x| \leq 1/2^n$.

As shown in the above algorithm, a SAR ADC requires:

1. An input voltage source Vin.
2. A reference voltage source Vref to normalize the input.
3. A DAC to convert the ith approximation xi to a voltage.
4. A Comparator to perform the function s(xi - x) by comparing the DAC's voltage with the input voltage.
5. A Register to store the output of the comparator and apply xi-1 - s(xi-1 - x)/2i.

An ADC that employs successive approximation relies on a comparator to eliminate potential voltage ranges before deciding on a single, definitive voltage range. The process of successive approximation involves making a continuous comparison of the input voltage to the output of an internal digital-to-analog converter (DAC, which is supplied by the value of the most recent approximation) in order to arrive at the most accurate possible approximation. Throughout the entirety of this procedure, a successive approximation register (SAR) will have a binary value representing the approximation saved there at each stage. When doing comparisons, the SAR makes use of a reference voltage, which is the largest signal that the ADC is expected to transform.

For instance, if the input voltage is 60 V and the reference voltage is 100 V, the 60 V is compared to 50 V (the reference value split by two) during the first clock cycle. This occurs because the reference voltage is multiplied by two. When the input is a "1" followed by zeros, and the voltage from the comparator is positive (or a "1") (since 60 V is larger than 50 V), this is the voltage that is output from the internal DAC. At this moment, the first binary digit in the most significant bit (MSB) has been changed to a 1. In the second clock cycle, the input voltage is compared to 75 V (being halfway between 100 and 50 V: This is the result of the internal DAC when its input is '11' followed by zeros). Given that 60 V is less than 75 V, the comparator output is now negative (or '0'). Therefore, a value of '0' is assigned to the second binary digit. The input voltage is compared with 62.5 V during the third clock cycle. 62.5 V is the output of the internal DAC when its input is "101" followed by zeros, therefore it is halfway between 50 V and 75 V. This is the value that is determined by the input voltage. Because 60 V is lower than 62.5 V, the output of the comparator is negative, which corresponds to the value '0'; hence, the third binary digit is also set to a '0'. The same thing happens during the fourth cycle of the clock, so this time the fourth digit is a '1' (since 60 V is higher than 56.25 V, which is the DAC output for '1001' followed by zeros). The binary representation of this outcome would be the number 1001. This process, which is very similar to a binary search, is sometimes referred to as bit-weighting conversion.

This particular converter type is a mid-rise converter, as indicated by the fact that the analogue value is rounded to the nearest binary number below. The conversion takes one clock cycle for each bit of resolution that is wanted since the approximations are performed one after the other rather than simultaneously. The sampling frequency must be multiplied by the number of bits of resolution that is wanted, and that number must be equal to the clock frequency. For instance, in order to sample audio at a frequency of 44.1 kHz with a resolution of 32 bits, a clock frequency of greater than 1.4 MHz is required. These kinds of ADCs have high resolutions and relatively broad ranges of operation. They are more difficult to construct than other types of designs.

4.10 DUAL-SLOPE ADC

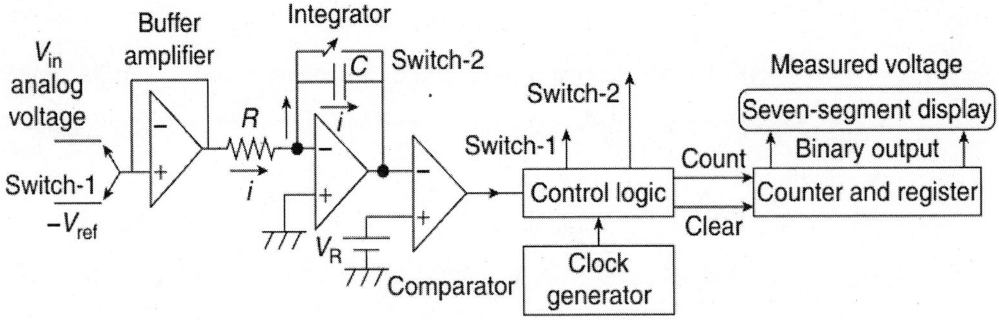

Fig :4.10 (a): Functional diagram of dual slope ADC

An integrating ADC, also known as a dual-slope ADC, is depicted in figure 4.10(a). This type of ADC connects the unknown input voltage to the input of an integrator and then allows the voltage to ramp for a predetermined amount of time (the run-up period). After then, a known reference voltage with the opposite polarity is given to the integrator, and this voltage is allowed to ramp up until the output of the integrator returns to zero (this step is known as the run-down time). It is possible to calculate the input voltage by taking into account the reference voltage, the constant run-up time period, and the recorded run-down time period. Since the measurement of run-down time is typically done in units of the converter's clock, integration times that are longer make it possible to achieve better resolutions. Similarly, the speed of the converter can be increased while maintaining the same level of resolution if the resolution is reduced. Because of its linearity and adaptability, converters of this type are utilised in the majority of digital voltmeters. Other variations on the concept may also be utilised.

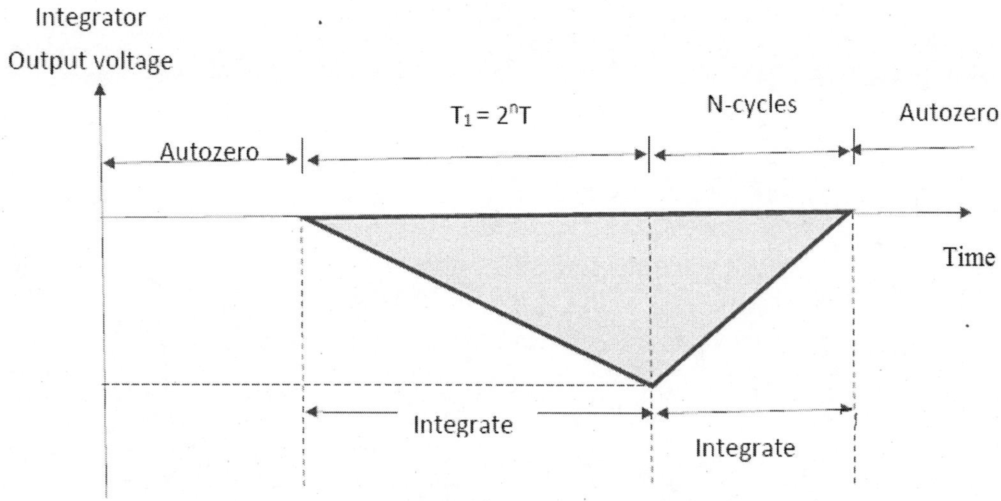

Fig.4.10 (b) : output waveform of dual slope ADC

During operation, the integrator is initialised to zero (by closing SW2), and then it is connected to the input (by turning SW1 to the "up" position) for a predetermined amount of time M counts of the clock (at a frequency of 1/t). At the conclusion of that period of time, it is connected to the reference voltage by sliding the switch to the down position (SW1), and the number of counts N that accrue before the integrator's output hits zero volts and the comparator's output changes is determined. Figure depicts the waveform of a dual slope analogue to digital converter. The equations of operation are therefore:

$$T_1 = t_2 - t_1 = \frac{2^n counts}{clock\ rate}$$

And

$$t_3 - t_4 = \frac{digital\ count\ N}{clock\ rate}$$

For an integrator, $\Delta V_0 = \frac{-1}{RC} V(\Delta t)$. The voltage Vo will be equal to V1 at the instant t2 and can be written as, $V_1 = \frac{-1}{RC} V_a(t_2 - t_1)$. The voltage V1 is also given by

$$V_1 = \frac{-1}{RC}(-V_R)(t_2 - t_3)$$

So, $V_a(t_2 - t_1) = (V_R)(t_3 - t_2)$. Putting the values of $(t_2 - t_1) = 2^n$ and $(t_3 - t_2) = N$, we get

$V_a(2^n) = (V_R)N$ Or, $V_a = (V_R)\left(\frac{N}{2^n}\right)$

SPECIFICATIONS FOR DAC/ADC

1. RESOLUTION: A converter's resolution refers to the smallest change in voltage that it is capable of producing at the output of the device. Resolution (in volts) = $(V_{FS})/(2^n-1)$ = 1 LSB increment

   Ex: An 8-bit D/A converter have 2^8-1=255 equal intervals. Hence the smallest change in output voltage is (1/255) of the full scale output range.

   An 8-bit DAC is said to have: 8 bit resolution

   : a resolution of 0.392 of full scale

   : a resolution of 1 part in 255

   In a similar manner, the resolution of an analog-to-digital converter is defined as the smallest change in analogue input that results in a change of one bit at the output.

   Ex: the input range of 8-bit A/D converter is divided into 255 intervals. So the resolution for a 10V input range is 39.22 mV = (10V/255)

2. LINEARITY: The linearity of an analog-to-digital (A/D) or digital-to-analog (D/A) converter is a crucial measurement of the accuracy of the converter. It informs us how close the output of the converter is to the ideal features of the converter.

3. GLITCHES (PARTICULARLY DAC): During the transition from one digital input to the next, such as from 0111 to 1000, it is possible that it will effectively go via 1111 or 0000, which will produce "unexpected" voltage for a small period of time. It may lead to difficulties in other areas.

4. ACCURACY: The maximum amount by which the actual converter output differs from the ideal converter output is referred to as absolute accuracy.

5. MONOTONIC: A DAC is said to be monotonic if its analogue output grows in proportion to an increase in the digital input. It is necessary for control-related applications.. If a DAC has to be monotonic, the error should be less than ±(1/2) LSB at each output level.

6. SETTLING TIME: The settling time is the dynamic parameter that is considered to be the most crucial. It is the amount of time that must pass for the output to stabilise within a predetermined band that is within (1/2) LSB of its final value after a code change has been made at the input. Because of the presence of internal parasitic capacitances and inductances, the switching time of the logic circuitry is directly affected by this. It can be anywhere between 100 ns and 10μ s.

7. STABILITY: The performance of the converter will differ depending on the temperature, its age, and the variations in the power source. Therefore, the stability is necessary.

UNIT SUMMARY

- The conversion of analogue signals to digital ones has its fundamental working principles broken down here. In order to produce a digital signal, an analogue signal must first be transformed into binary words. Displaying the binary data and making it accessible for additional use is both possible.
- Circuits that convert analogue signals to digital are divided into two categories:
  (a) Direct type ADCs such as counter type ADCs, SA type ADCs, and parallel comparator (Flash) ADCs(b) ADCs of the Integrating type, such as single slope type and dual slope type ADCs
- The operation of multiple different kinds of DAC circuits is broken down and illustrated using the appropriate diagrams.
- For the purpose of converting digital data, which represents binary bits, into analogue data, a DAC circuit is utilised.
- In order to transform binary (digital) signals into analogue signals, operational amplifiers and resistor networks are utilised in (a) binary weighted DACs, (b) R-2R

ladder network DACs, and (c) inverted R-2R ladder type DACs. These DACs are classified according to the sort of ladder network they utilise. These circuits are illustrated with nifty circuit diagrams in the explanations. Diffused, Epitaxial, pinched, thin film resistors are most commonly used one.

Multiple Choice Questions

1. Percentage resolution of an ADC with binary digits of n = 8
   (a) 0.391
   (b) 6.25
   (c) 0.0244
   (d) 2328×10^{-9}

   [Ans. (a)]

2. Percentage resolution of an ADC with binary digits of n = 4
   (a) 0.391
   (b) 6.25
   (c) 0.0244
   (d) 2328×10^{-9}

   [Ans. (b)]

3. Disadvantage of a parallel comparator type ADC
   (a) Number of bits increases as complexity of signal increases
   (b) Divider network needs a large resistive network
   (c) Requirement of comparators doubles with increase of every bit
   (d) All of these

   [Ans. (c)]

4. Fastest ADC
   (a) Dual slope ADC
   (b) Flash type ADC
   (c) Counter type ADC
   (d) Single slope type ADC

   [Ans. (b)]

5. Low-speed ADC
   (a) Dual slope ADC

(b) Flash type ADC

(c) Counter type ADC

(d) SA type ADC

[Ans. (d)]

6. Advantage of a dual slope type ADC

(a) Accurate measurement of slowly varying signals

(b) Fast conversion process

(c) Long conversion time

(d) Excellent noise rejection

[Ans. (a)]

Answers of Multiple Choice Questions

Short and Long Answer Type Questions

1. Define the specifications of ADC?
2. Explain about Dual Slope ADC.
3. Derive the output voltage expression of a 4 bit R-2R Ladder converter.
4. Explain about the operation of Successive Approximation ADC
5. Draw and explain about Flash Converter

REFERENCES AND SUGGESTED READINGS

1. Coughlin, R. F. and Frederick F. Driscoll (2001). Operational Amplifiers and Linear Integrated Circuits, Sixth Edition. Upper Saddle River, NJ: Prentice Hall.
2. Dailey, Denton J. (1989) Operational Amplifiers and Linear Integrated Circuits: Theory and Applications. New York: The McGraw-Hill Companies.
3. Franco, Sergio (2002). Design with Operational Amplifiers and Analog Integrated Circuits. New York: The McGraw-Hill Companies.
4. Gayakwad, Ramakant A. (1999). Op-Amps and Linear Integrated Circuits. Fourth edition. Upper Saddle River, NJ: Prentice Hall Inc.
5. Jacob, J. Michael (1996) Applications and Design with Analog Integrated Circuits. Upper Saddle River, NJ: Prentice Hall Inc.
6. *Ramakanth A. Gayakwad* (2000),Op-Amps and Linear Integrated Circuits -, *Prentice-Hall, 4th edition.*

7. S.Salivahanan,V S Kanchana Bhaskaran(2015),Linear Integrated Circuits, Second Edition, Tata McGrawa Hill Edition.
8. B.Visweswara Rao(2015), Linear Integrated Circuits, First Edition, Pearson India Education Services Pvt. Ltd

Dynamic QR Code for Further Reading

5 Waveform Generators and Special Function IC's

UNIT SPECIFICS

Through this unit we have discussed the following aspects:

- *Different types of Waveform Generators;*
- *Different types of Special Integrated circuits;*
- *Working principles of 555 timer;*
- *Discussion on Audio, Video and other types of Amplifiers;*
- *Discussion on Voltage to Frequency and Frequency to voltage converters*

RATIONALE

In electronic circuits, operational amplifiers, or op-amps, are frequently employed to produce a variety of waveforms. The majority of analogue and digital devices require one or more periodic waveforms for a variety of functions, including timing and control, respectively. Other kinds of evolution in the design of operational amplifiers include the generation of waveforms most typically utilised, such as sinusoidal, square, and triangular waveforms. Their usage made it possible to develop oscillators, which are able to generate repetitive waveforms of a constant frequency and amplitude without the requirement for any other signal to be present. These oscillators can be used in a variety of applications. The electronic circuits that are utilised in order to generate various waveforms are referred to as oscillators, function generators, and waveform generators, respectively. The topics of sinusoidal oscillators and non-sinusoidal oscillators, such as multivibrators and applications of timer IC 555, are covered in this chapter. In addition to that, the monolithic ICL8038 function generator and the uses of it are discussed. A separate section on 555 timer is introduced. Audio Amplifier, Video Amplifier, Isolation amplifiers are also explained to the end user.

PRE-REQUISITES

Electronic devices and circuits, Circuit theory

UNIT OUTCOMES

List of outcomes of this unit is as follows:

U5-O1: *Describe about 555 timer*

U5-O2: *Explain the different wave form generators*

U5-O3: *Explain Audi amplifiers, video amplifier*

U5-O4: *Discuss about voltage to frequency and frequency to voltage converters*

U5-05: *Develop the concept of voltage regulator*

Unit-5 Outcomes	**EXPECTED MAPPING WITH COURSE OUTCOMES** (*1-Weak Correlation; 2- Medium Correlation; 3-Strong Correlation*)					
	CO-1	**CO-2**	**CO-3**	**CO-4**	**CO-5**	**CO-6**
U5-01	-	2	2	2	1	3
U5-02	1	3	2	3	2	3
U5-03	-	3	3	2	2	3
U5-04	1	2	3	2	3	3
U5-05	2	2	2	2	1	3

5.1 SQUARE WAVEFORM GENERATORS.

In contrast to sine wave oscillators, square wave outputs are generated when the op-amp is forced to operate in the saturated region. That is, the output of the op-amp is forced to swing repetitively between positive saturation $+V_{sat}$ ($\approx +V_{CC}$) and negative saturation $- V_{sat}$ ($\approx +V_{EE}$), resulting in the square-wave output.

Figure 5.1(a) depicts one example of such a circuit. This type of square wave generator is sometimes referred to as a free-running multi-vibrator or a stable multi-vibrator. When the differential voltage V_{id} is either positive or negative, the output of the op-amp in this circuit will either be in a state of positive or negative saturation, depending on which of those two states it is in.

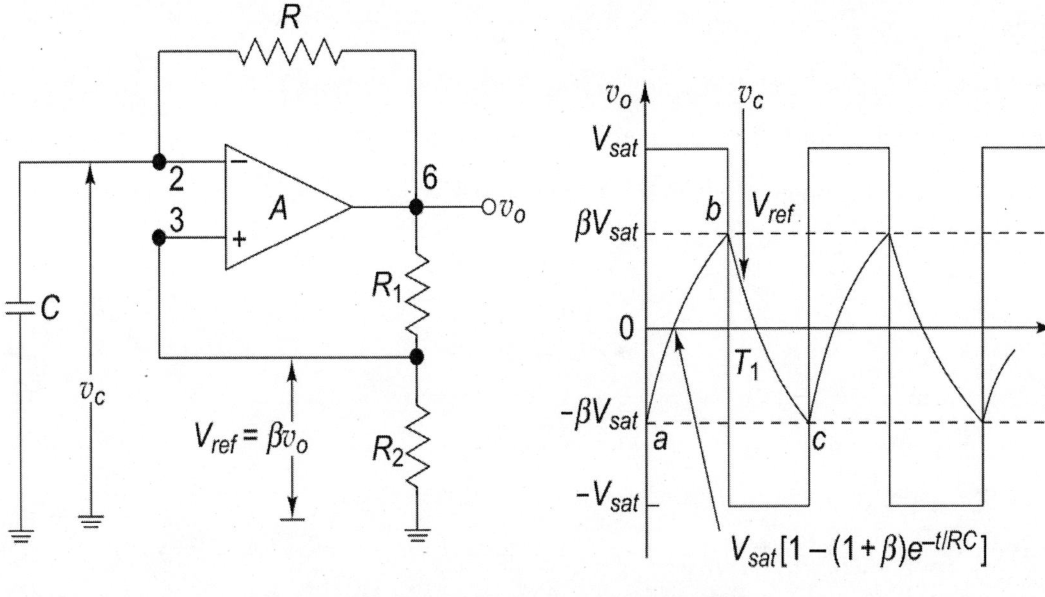

Fig.5.1.Square wave form generator

Assume that the voltage across capacitor C is 0 volts at the moment that the dc supply voltages $+V_{CC}$ and $-V_{EE}$ are applied. This will ensure that the calculations are accurate. This indicates that there is no voltage present at the inverting terminal at the beginning. However, at the same time, the voltage V_1 at the non-inverting terminal is a very small finite value that is a function of the output offset voltage V_{OOT} as well as the values of the resistors R_1 and R_2. As a result, the differential input voltage V_{id} is the same as the voltage measured at the terminal that does not invert the signal. Voltage V_1 will start to drive the operational amplifier into saturation, despite the fact that it is quite low. Take, for instance, the scenario in which the output offset voltage V_{OOT} is positive and, as a consequence, the voltage V_1 is also positive. Because initially the capacitor C functions as a short circuit, the gain of the operational amplifier is very high (A). As a result, V_1 drives the output of the operational amplifier to its positive saturation, which is denoted by $+V_{sat}$. When the voltage at the output of the operational amplifier is set to $+V_{sat}$, the capacitor C begins charging towards $+V_{sat}$ through the resistance provided by the resistor R. However, once the voltage across capacitor C reaches a point where it is marginally more positive than V_1, the output of the operational amplifier is compelled to flip to a negative saturation, shown by the symbol $-V_{sat}$. With the op-amp's output voltage at negative saturation, $-V_{sat}$, the voltage v_1 across R_1 is also negative, since

$$V1 = \frac{R1}{R1 + R2}(-Vsat)$$

Therefore, the net differential voltage $V_{id}=V_1-V_2$ is negative, which maintains the output of the op amp in negative saturation. The output will continue to be in negative saturation until the capacitor C discharges and then recharges to a negative voltage that is slightly higher than $-V_1$. However, as soon as the capacitor's voltage V_2 becomes more negative than $-V_1$, the net differential voltage V_{id} will become positive, which will drive the output of the op amp back to its positive saturation $+V_{sat}$. This completes one cycle. With output at $+V_{sat}$, voltage V1at the non-inverting input is

$$V1 = \frac{R1}{R1 + R2}(+Vsat)$$

The time period T of the output waveform is given by

$$T = 2RC\ln\left(\frac{2R1 + R2}{R2}\right)$$

$$f0 = \frac{1}{2RC\ln\left(\frac{2R1 + R2}{R2}\right)}$$

Above equation indicates that the frequency of the output f_0 is not only a function of the RC time constant but also of the relationship between R_1andR_2. For example, if $R_2=1.16R_1$, Equation becomes $f0 = \frac{1}{2RC}$

5.2 TRIANGULAR WAVE GENERATOR

Remember that if a square wave is fed into the integrator, a triangle pattern will result. This means that all you have to do to create a triangular wave generator is connect an integrator to the square wave generator. Figure5.2(a) depicts the constructed circuit. This design requires at least five resistors and two capacitors. Also required is a dual op-amp.

The frequency of the square wave and the triangle wave is identical. For constant values of R_1, R_2, and C, the frequency of the square wave and the frequency of the triangle wave depend on the resistance R. Depending on the value of R, the triangle wave frequency increases or decreases as R changes. Even though the amplitude of a square wave remains constant (V_{sat}), the amplitude of a triangular wave decreases as its frequency increases, and vice versa. The square wave is input into the A_2 processor, and the triangle wave is output.

To generate a triangular wave from A_2, $5R_3C_2$ must be greater than T/2, where T is the duration of the input square wave. The correct answer for R_3C_2 is T. To obtain a table triangular wave, it may be necessary to shunt capacitor C_2 with resistance $R_4=10R_3$ and connect an offset voltage-compensating network to the non-inverting end of A_2. The diagram below depicts a triangle wave generator with fewer components. The generator contains an A_1 comparison and an A_2 adder. Continuously, the comparator A_1 compares the voltage at point P to the inverted 0V input.When the voltage at P falls slightly below 0V or rises slightly above 0V, the output of A_1 reaches negative or positive saturation, respectively.

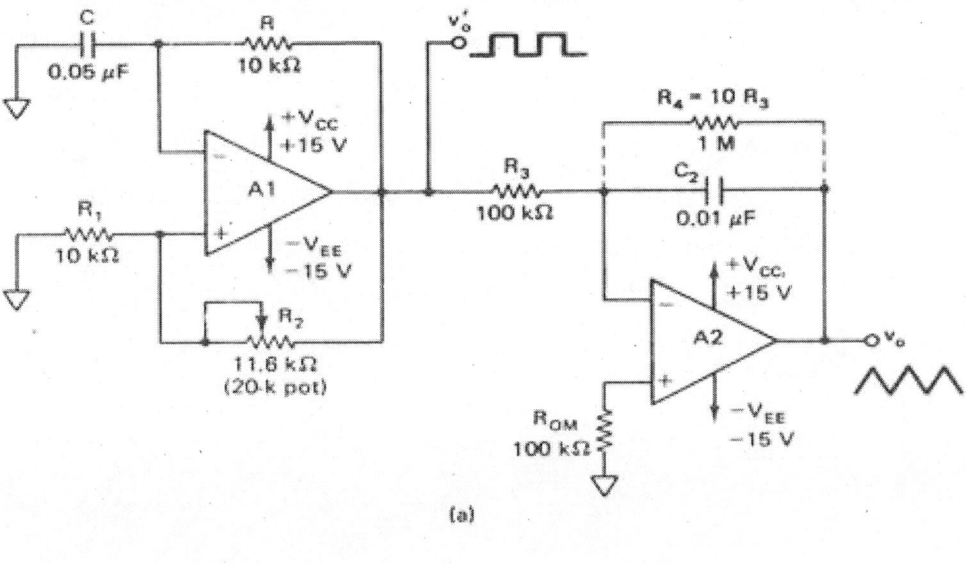

(a)

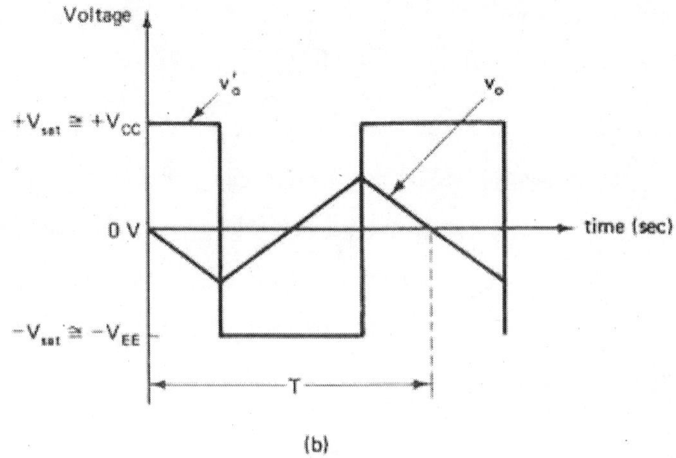

Fig.5.2.: (a) Triangular Wave Generator (b) Output Wave Form

Let's start by giving the output of A a positive saturation of +V (+Vc) so that we can provide a clearer explanation of how the circuit operates. This +V provides the integrator A_2, which may be found over here, with an input signal. Given that it is the output of A_2, the ramp that it constitutes will be one that moves in the opposite direction. Because of this, one end of the voltage-divider that is comprised of R_2-R_3 corresponds to the positive saturation voltage +V of A, while the other end corresponds to the negative-going ramp of A2. When the ramp travelling in the other direction reaches a certain value, which we'll refer to as V Ramp, the point P will be placed just a hair's breadth below 0 V. As a direct consequence of this, the output of A1 will go from having positive saturation to having negative saturation. This suggests that the output of A_2 will now begin to go positively, as opposed to continuing its previous trend of moving negatively as it had been doing previously. The amount of output that is generated by A_2 will continue to increase up until the point where it reaches Positive Value.As a consequence of this, the output of A has been moved back to the positive saturation level +V at this same instant because point P is now only slightly higher than 0V. After that, the procedure is carried out again. The waveform of the output is exactly the same as what is shown in the illustration.

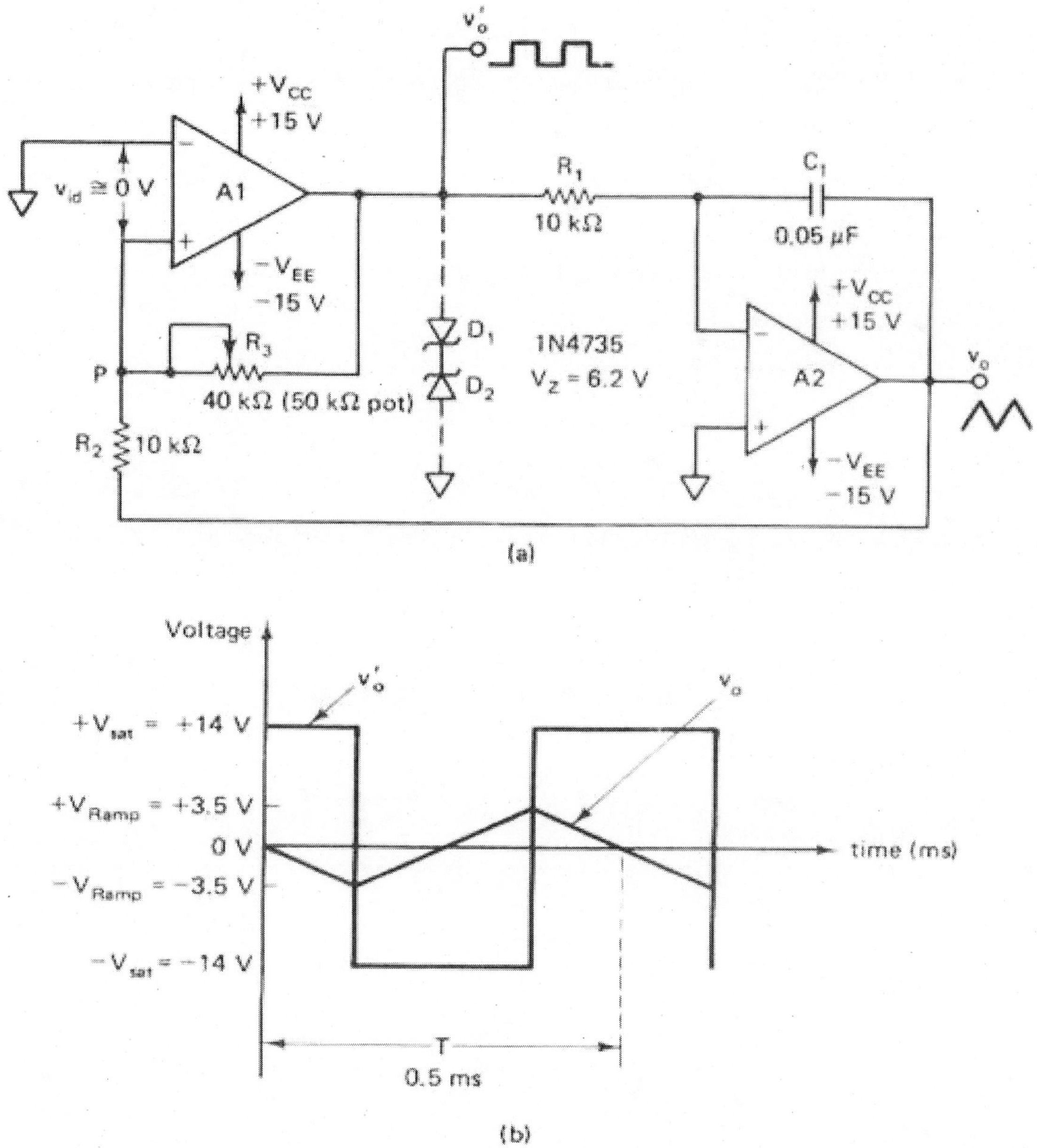

Fig.5.3: Triangular Wave Generator with zener diodes(a) Circuit (b) Input and Output Waveform

The square wave and the triangle wave have the same frequencies throughout their entirety. The amplitude of the square wave is a function of the voltages provided by the dc supply. On the other hand, one can get the correct amplitude by connecting the proper zeners to the output of amplifier A1.

The following formula can be used to calculate both the amplitude and the frequency of the triangle wave: Figure (b) shows that when the output of comparator A_1 is +V, the output of integrator A_2 begins a steady decline that continues until it reaches a value that is negative V_{rn}.At this point, the voltage at the output of A_1 will change from +V to -V. Immediately prior to the occurrence of this switching, the voltage at point P (+input) is 0V. Therefore, the -V $_{Ramp}$ needs to be built over R_2, and the $+V_{sat}$ needs to be created across R_3. That is,

$$\frac{-VRamp}{R2} = -\frac{+Vsat}{R3}$$

$$-VRamp = -\frac{R2}{R3}(+Vsat)$$

Similarly, +Vj,, the output voltage of A2 at which the output of A_1 switches from –V to +V, is given by,$+VRamp = -\frac{R2}{R3}(-Vsat)$.The peak-to-peak (pp) output amplitude of the triangular wave is

$$V0(pp) = +VRamp - (-VRamp)$$

$$V0(pp) = 2\frac{R2}{R3}(Vsat)$$

Above equation indicates that the amplitude of the triangular wave decreases with an increase in R_3. The time it takes for the output waveform to swing from - to + (or from +V Ramp to -V Ramp) is equal to half the time period T/2.This time can be calculated from the integrator output equation.

$$V0(pp) = -\frac{1}{R1C1}\int_0^{T/2}(-Vsat)\,dt$$

The frequency of oscillation then is

$$f0 = \frac{R3}{4R1C1R2}$$

5.3 SAWTOOTH WAVEGENERATOR

The rise time of a triangle wave is always equal to its fall time, whereas the sawtooth wave's rise time is never equal to its fall time. This is the primary distinction between the triangular and sawtooth waveforms. Therefore, the amount of time necessary for the triangle wave to swing from $-V_{ramp}$ to $+V_{ramp}$ is the same as the amount of time necessary for it to swing from $+V_{ramp}$ to $-V_{ramp}$. The sawtooth waveform, on the other hand, has different amounts of time spent rising and falling. That is, it may go up in a positive direction a great deal faster than it goes down in a negative direction, or vice versa.

By injecting a variable dc voltage into the non-inverting terminal of the integrator A_2, the triangle wave generator shown in Figure (a) can be turned into a sawtooth wave generator. Using the potentiometer and connecting it to the $+V_{CC}$ and $-V_{EE}$ as illustrated in Figure (a) is one method that can be utilised to achieve this goal.

The output of A_2 has a particular dc level added to it, and this level is determined by the setting of R_4. Let's say for the sake of argument that the output of A_1 is a square wave, and that the potentiometer R_4 has been tuned to a particular dc level.

This indicates that the output of A_2 will be a triangle wave, riding on some dc level that is determined by the value of R_4 in the amplifier. The polarity and amplitude of this dc level will serve as the primary determinants of the duty cycle of the square wave. The output of A_2 will then be sawtooth-shaped if the duty cycle is less than 50%.

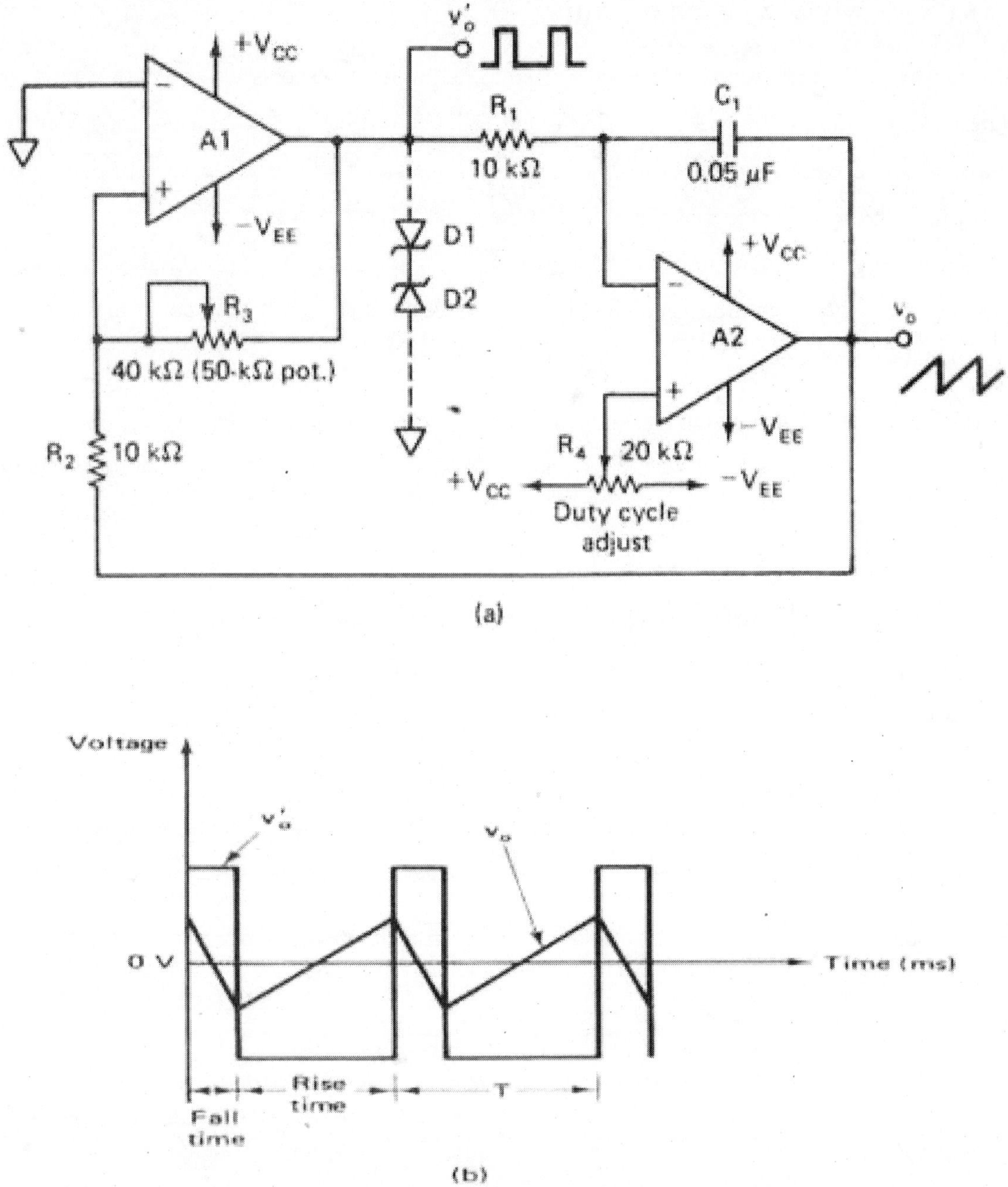

Fig.5.4: Saw tooth Wave generator (a) Circuit (b) Output Waveform With the wiper at the centre of R4, the output of A2 is a triangular wave.

For any other position of R_4 wiper, the output is a sawtooth waveform. Specifically as the R4 wiper is moved toward— V, the rise-time of the sawtooth wave becomes longer than the fall time. On the other hand, as the wiper is moved toward $+V_{cc}$, the fall time becomes longer than the rise-time. Also, the frequency of the sawtooth wave decreases as $R_4$is adjusted toward + V or — V_{EE}. However, the amplitude of the sawtooth wave is independent of theR_4setting.

5.4 FUNCTION GENERATOR IC 8038:

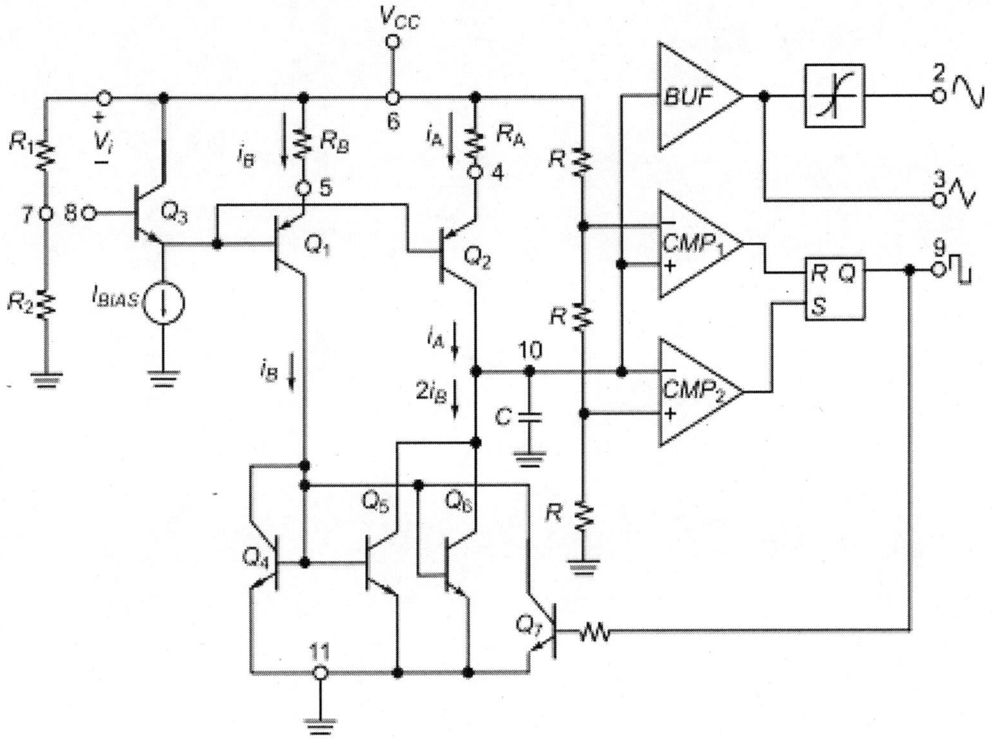

Fig.5.5 8038 IC block diagram and waveforms

It is made up of two different current sources, two different comparators, two different buffers, one FF, and a sine wave converter.The block diagram of the functional components of the Function Generator IC 8038 can be seen in figure 5.5.Figure 5.5 illustrates the various waveforms that can be expected from the IC 8038's output.

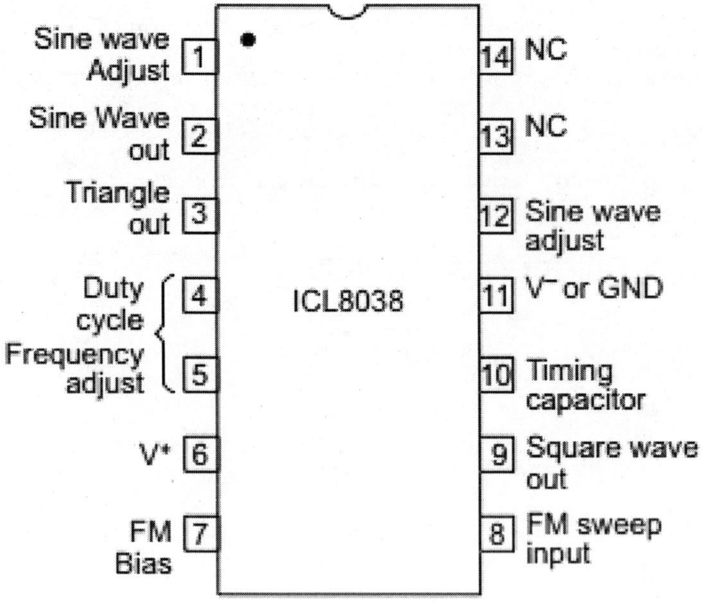

Pin description:

1. Pin 1 & Pin 12: Sine wave adjusts:

   By changing the 100K pots connected between pins 12 and 11 and between pins 1 and 6, the distortion in the sine wave output can be decreased.

2. Pin 2 Sine Wave Output:

   Sine wave output is available at this pin. The amplitude of this sine wave is 0.22 Vcc. Where $\pm 5V \leq Vcc \leq \pm 15$ V.

3. Pin 3 Triangular Wave output:

   Triangular wave is available at this pin. The amplitude of the triangular wave is 0.33Vcc.Where $\pm 5V \leq Vcc \leq \pm 15$ V.

4. Pin 4 & Pin 5 Duty cycle / Frequency adjust:

   The external resistors linked from Vcc to pin 4 are used to modify the symmetry of all output waveforms and the 50% duty cycle for the square wave output. The frequency of the output wave forms will be controlled by these external resistors and capacitors at pin 10.

5. Pin 6 + Vcc:

Positive supply voltage the value of which is between 10 & 30V is applied to this pin.

6. Pin 7 : FM Bias:

   This pin along with pin no8 is used to TEST the IC 8038.

7. Pin 9 : Square Wave Output:

   This pin has a square wave output available. Because it has an open collector output, this pin can be connected to various power supply voltages via a load. The square wave output can be produced extremely effectively with this method.

8. Pin 10 : Timing Capacitors:

   The external capacitor C connected to this pin will decide the output frequency along with the resistors connected to pin 4 & 5.

9. Pin 11 : -VEE or Ground:

   If a single polarity supply is to be used then this pin is connected to supply ground & if (±) supply voltages are to be used then (-) supply is connected to this pin.

10. Pin 13 & Pin 14: NC (No Connection)

Important features of IC 8038:

1. All the outputs are simultaneously available.
2. Frequency range : 0.001Hz to 500kHz
3. Low distortion in the output wave forms.
4. Low frequency drifts due to change in temperature.
5. Easy to use.

Parameters:

(i) Frequency of the output wave form:

The output frequency dependent on the values of resistors R1 & R2 along with the external capacitor C connected at pin 10.If $R_A = R_B = R$ & if R_C is adjusted for 50% duty cycle then $f_0 = 0.3/R_C$; $R_A = R_1$, $R_B = R_3$, $R_C = R_2$.

(ii) Duty cycle / Frequency Adjust : (Pin 4 & 5):

Duty cycle as well as the frequency of the output wave form can be adjusted by external resistors at pin 4 & 5.The values of resistors R_A & R_B connected between V_{cc} pin 4 & 5 respectively along with the capacitor connected at pin 10 decide the frequency of the wave form. The values of R_A & R_B should be in the range of 1kΩ to 1MΩ.

(iii)FM Bias:

- The FM Bias input (pin7) corresponds to the junction of resistors R_1 & R_2.
- The voltage Vin is the voltage between V_{cc} & pin8 and it decides the output frequency.
- The output frequency is proportional to V_{in} as given by the following expression.

For $R_A = R_B$ (50% duty cycle). $f_0= 5\ V_{in}/CR_AV_{cc}$; where C is the timing capacitor.With pin 7 & 8 connected to each other the output frequency is given by $f_0= 0.3/R_C$ where $R = R_A = R_B$ for 50% duty cycle.

This is because M Sweep input (pin 8):

$V_{in}=R_1V_{cc}/R_1+R_2$

- This input should be connected to pin 7, if we want a constant output frequency. But if the Output frequency is supposed to vary, then a variable dc voltage should be applied to this pin.
- The voltage between Vcc & pin 8 is called Vin and it decides the output frequency as,

$f_0=1.5\ V_{in}/CR_AV_{cc}$

A potentiometer can be connected to this pin to obtain the required variable voltage required to change the output frequency.

5.5 INTRODUCTION TO 555 TIMER:

The 555 timer is one of the most adaptable linear integrated circuits. Mono-stable and astable multivibrators, dc-dc converters, waveform generators, analogue frequency metres and tachometers, temperature measurement and control, infrared transmitters, burglar and toxic gas alarms, voltage regulators, electric eyes, and many other devices are examples of these applications.A monolithic timing circuit, the 555 can generate precise and extremely stable time delays or oscillations. Either as an astable (free running) multivibrator or a monostable (one-shot) multivibrator, the timer essentially runs in one of these two modes. The component is offered as a 14-pin DIP, an 8-pin micro DIP, or an 8-pin metal container.

The SE555 is designed for the operating temperature range from -55 $^{\circ C}$to + 125$^{\circ C}$, while the NE555 operates over a temperature range of 0° to +70$^{\circ C}$. The important features of the 555 timer are these: it operates on +5 to + 18 V supply voltage in both free-running (astable) and one- shot (monostable) modes; it has an adjustable duty cycle; timing is from microseconds through hours; it has a high current output; it can source or sink 200 mA; the output can drive TTL and has a temperature stability of 50 parts per million (ppm) per degree Celsius change in temperature, or equivalently 0.005%/$^{\circ C}$. Like general-purpose op-amps, the 555 timer is reliable, easy to us, and low cost.

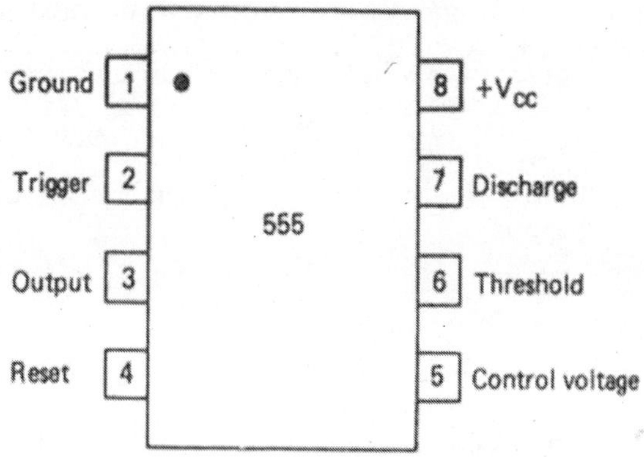

Fig.5.6.Pin Diagram of 555 Timer

Pin 1: Ground.

All voltages are measured with respect to this terminal.

Pin 2: Trigger.

The size of the external trigger pulse applied to this pin determines the timer's output. If the voltage at this pin is more than 2/3 of V_{CC}, the output is low. However, the output of the comparator 2 goes low when a negative pulse with an amplitude greater than 1/3$\cdot V_{CC}$ is supplied to this pin, which switches the output of the timer to high. As long as the trigger terminal is kept at a low voltage, the output stays high.

Pin 3: Output.

It is possible to connect a load to the output terminal in one of two ways: either between pin 3 and ground (pin 1) or between pin 3 and supply voltage + V_{CC} (pin 8). The sink current, which flows via the load connected between pin 3 and + V_{CC} into the output terminal when the output is low, is the load current.

However, when the output is low, there is no current flowing through the grounded load. Due to this, the load connected between pin 3 and + V_{CC} is referred to as the ordinarily on load, whereas the load linked between pin 3 and ground is referred to as the normally off load.

On the other hand, when the output is high, there is no current flowing through the load that is ordinarily connected between pin 3 and + V_{CC}. The output terminal, however, provides current to the ordinarily idle load. The source current is what is known as. 200 mA is the maximum amount of sink or source current.

Fig : Pin diagram of 555Timer

Pin 4: Reset.

When a negative pulse is applied to this pin, the 555 timer can be reset (disabled). To prevent any possibility of false triggering while the reset function is not in use, the reset terminal should be connected to + V_{CC}.

Pin 5: Control voltage.

Both the threshold and the trigger voltage are altered by an external voltage provided to this terminal. In other words, the output waveform's pulse width can be adjusted by applying a voltage to this pin or by connecting a pot between it and ground. When not used, the control pin should be bypassed to ground with a 0.01-µF capacitor to prevent any noise problems.

Pin 6: Threshold.

Comparator 1's non-inverting input terminal, which tracks the voltage across the external capacitor, is located here. When the voltage at this pin reaches the threshold voltage of 2/3 V, comparator 1's output goes high, switching the output of the timer to low.

Pin 7: Discharge.

According to Figure 5.6, this pin is internally connected to transistor Q_1's collector. Q_1 is off and functions as an open circuit to the external capacitor C connected across it when the output is high. The external capacitor C is shorted out to ground when the output is low because Q_1 is saturated and functions as a short circuit.

Pin 8: + V_{CC}.

The supply voltage of +5 V to +18 is applied to this pin with respect to ground (pin1).

FUNCTIONAL BLOCK DIAGRAM OF 555 TIMER:

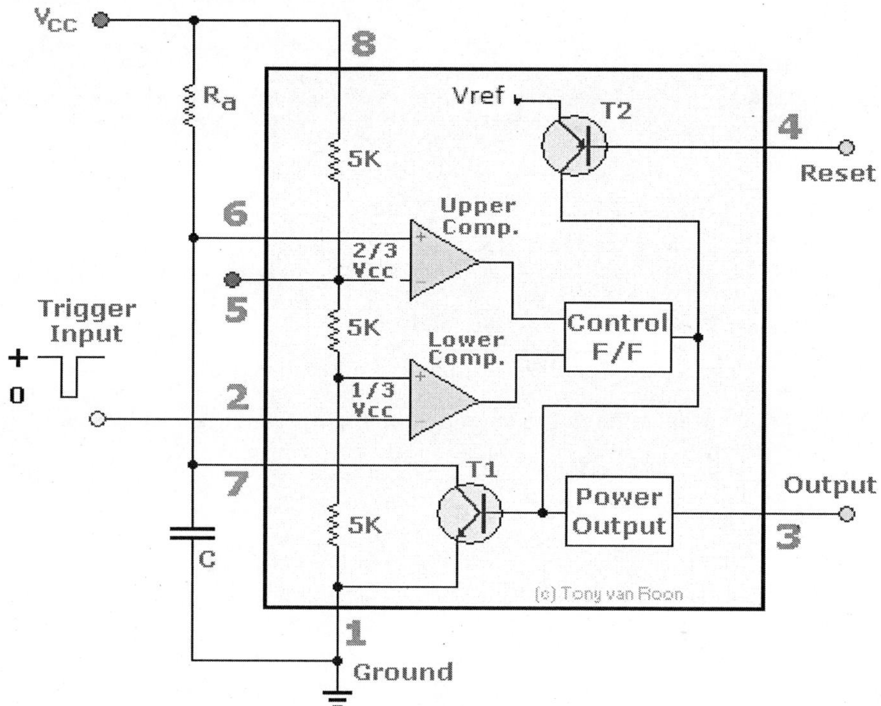

Fig 5.7. Block diagram of 555 timer

5.6 FREQUENCY TO VOLTAGE CONVERTORS (F-V)

F-V convertors finds applications in Tachometer in motor speed control Rotational speed measurement. Ideal characteristics of V-F converter and F-V converter is shown in figure 5.10.1. F-V convertor produces an output voltage whose amplitude is a function of input signal frequency. V0=$k_f f_i$ k_f is sensitivity of F-V convertor. Frequency to voltage converter using is shown in figure 5.8.

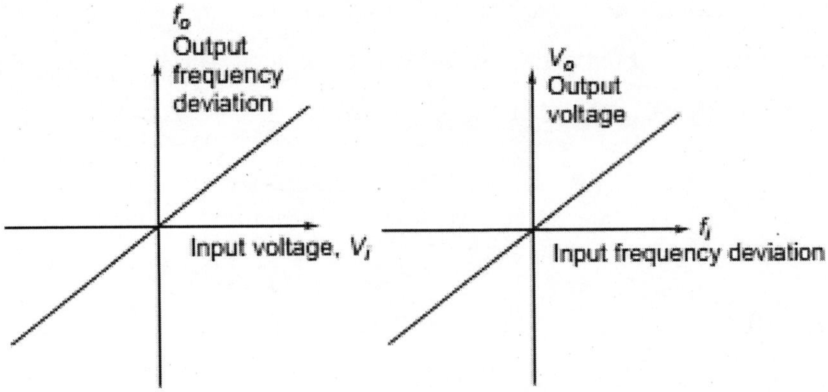

Fig.5.8 Characteristics of voltage to frequency and frequency to voltage converters

In essence, it acts as an FM discriminator. Comparator A receives input frequency. R is a feedback element, or resistor. Charge balance is made possible by capacitor C_i. Input signal is conditioned by a high pass network. Comparator COMP initiates a one-shot multivibrator with a threshold of 7.5V for a negative spike of V_{01}. When the multivibrator's output closes the switch SW for a brief period of time TH, voltage V_o builds up and is continuously injected through R. This process continues until the current leaving the opamp's summation input is equal to the voltage continually injected via R. Figure 5.9(a) displays the input-output characteristics of a frequency-to-voltage converter utilising a VFC32.

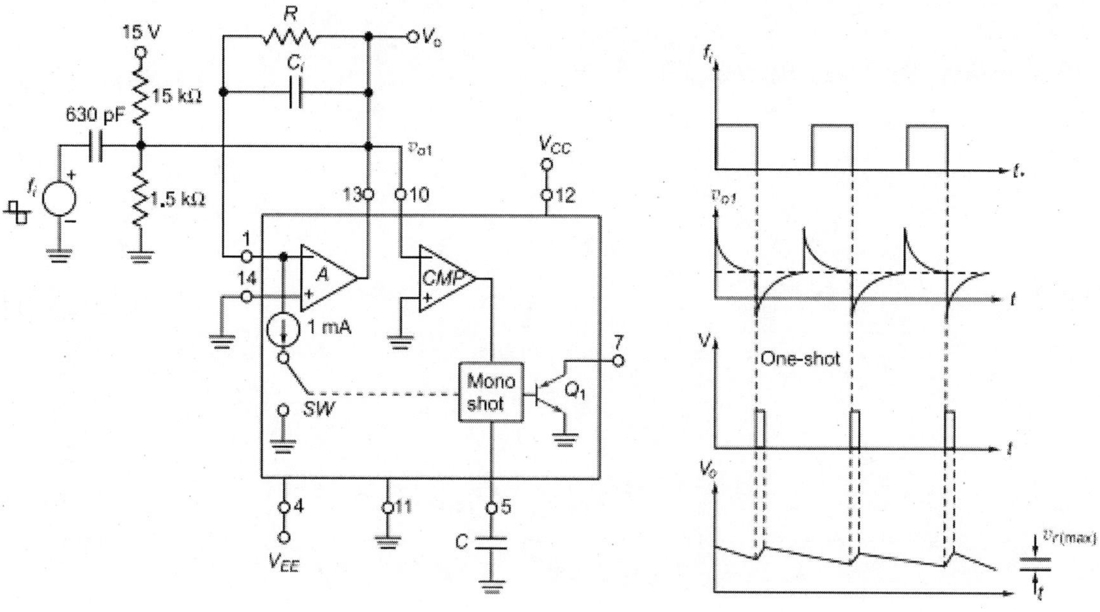

Fig.5.9(a) Voltage to Frequency Converter (b) Waveforms

$$V_o=10^{-3} *TH *R*fi \text{ as } TH =7.5\, C /1X10^{-3}$$

$$\text{Ripple Voltage, } V_{r(max)} =7.5\, C /C_i$$

VOLTAGE TO FREQUENCY CONVERTOR

The charge balancing technique is the fundamental operating principle; the process of charging and discharging produces frequency proportional to input signal F0=k Vi. The Operation is described in the next section. Op-amp Input Vi is transformed by A into current Ii = Vi/R at the summing junction.When switch SW is open, current enters the capacitor and charges it. Node voltage Vol also produces a ramp-down when this happens.One shot multivibrator receives a triggering signal from CMP when V01 = 0, which closes switch SW and activates transistor Q for time TH. Figure 5.10.4 illustrates the voltage-frequency characteristics of the VF32 and its input-output characteristics.The threshold of mono shot = 7.5 V and T_H= 7.5 C/10^{-3}. During T_H , V01 ramps upward by amount ΔV01=(1mA-Ii) T_H /C_i

Time duration T_L for vo1 to return to 0 is $T_L = C\Delta V_{01}/I_i$

$T_L+T_H = 1mA\ T_H/I_i = T$. and $F_0=V_i/7.5\ RC$

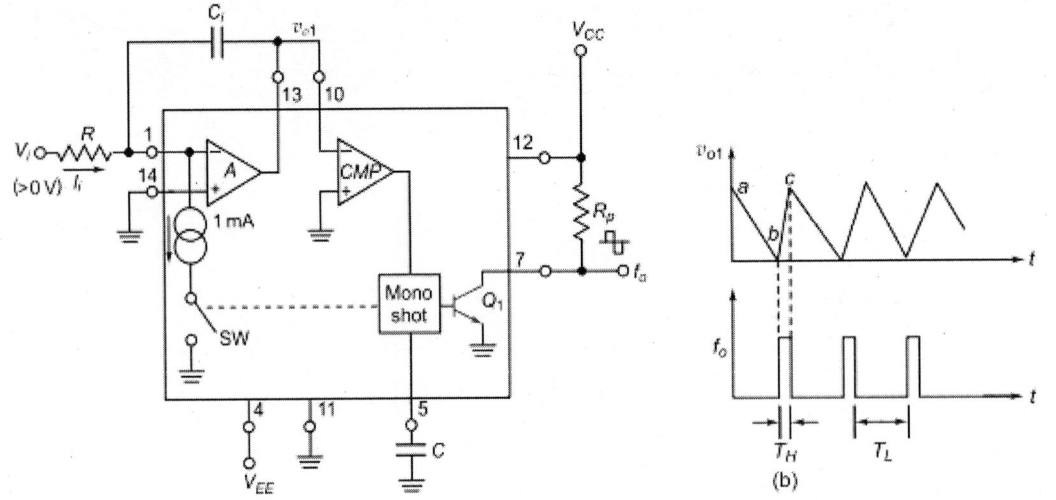

Fig.5.10. Voltage to Frequency Converter circuit diagram and waveforms

SWITCHED CAPACITOR FILTER ICS

Some of the Switched capacitor filter ICs is MF 5, MF10 and MF100

MF10:

The MF10 contains two of the second-order universal filter sections found in the MF5. Therefore with MF10, two second order filters or one fourth-order filter can be built. As the MF5 and MF10

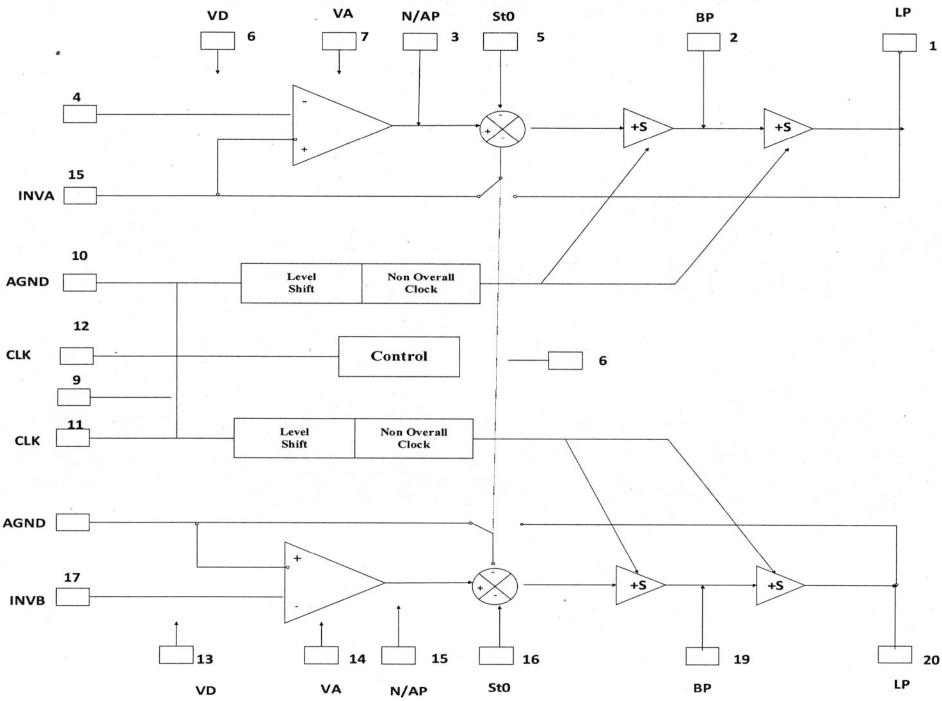

Fig.5.11. Switched Capacitor Filter

have similar filter sections, the design procedure for them is same. Figure 5.11.shown below is the Switched Capacitor Filter MF10

5.7 POWER AUDIO AMPLIFIER IC LM 380

Introduction:

In essence, voltage amplifiers that provide their loads with higher amplifier signal voltages are small signal amplifiers. huge signal or power amplifiers, on the other hand, deliver a huge signal current to current-operated loads like speakers and motors. The functional block diagram of an audio power amplifier is given in Figure 5.12 below.

However, the current required by the amplifier in audio applications is substantially larger than what is provided by general-purpose op-amps. As a result, the output of general purpose op-amps cannot be used to directly drive loads like speakers and motors that require large currents. Figures 5.13 which illustrate the pin diagram and block diagram of the LM380 power amplifier, respectively. The following is done to handle it.

1. To use discrete or monolithic power transistors called power boosters at the output of the op-amp.
2. To use specialized ICs designed as power amplifiers like LM 380.

Features of LM380:

1. Internally fixed gain of 50 (34dB)

2. Output is automatically self centering to one half of the supply voltage.

3. Output is short circuit proof with internal thermal limiting.

4. Input stage allows the input to be ground referenced or ac

5. Wide supply voltage range (5 to 22V).

6. High peak current capability.

7. High impedance.

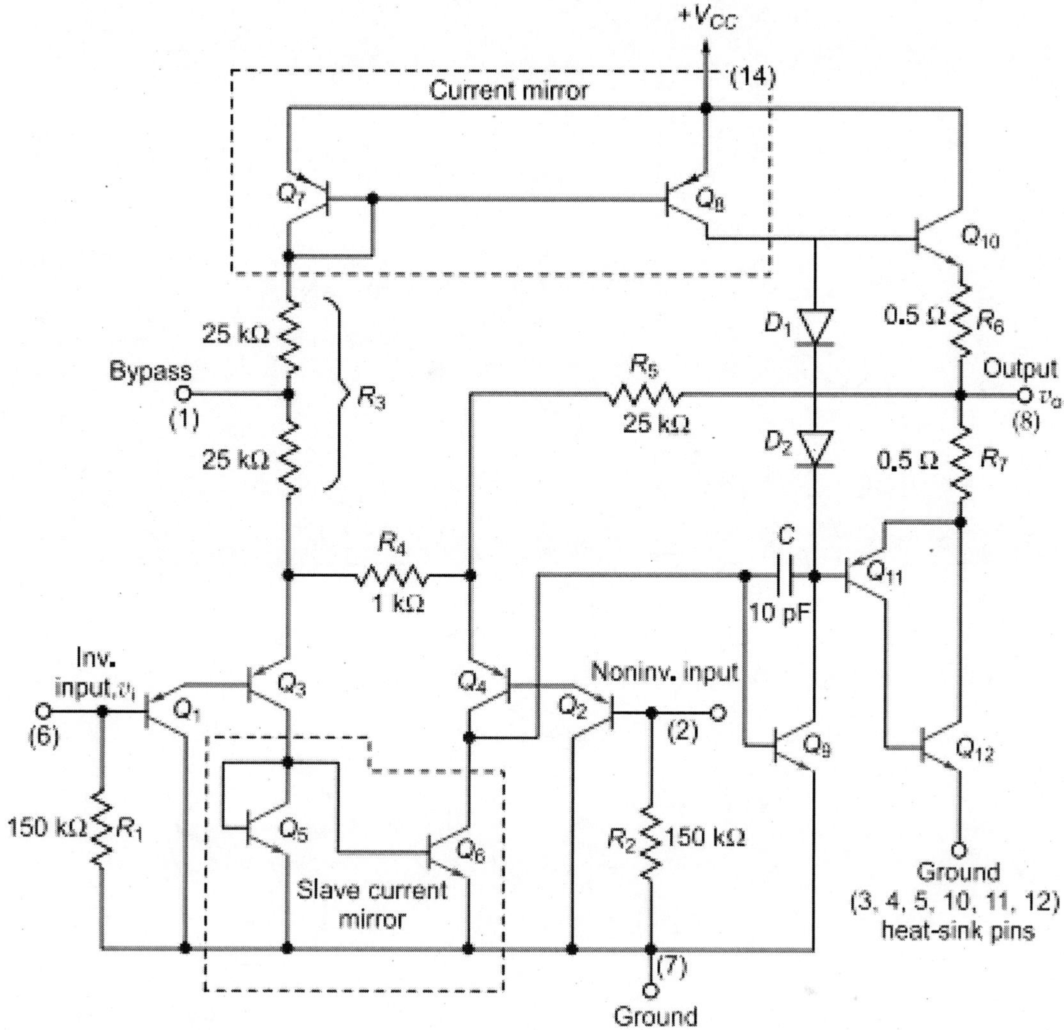

Fig.5.12. LM380 Circuit diagram

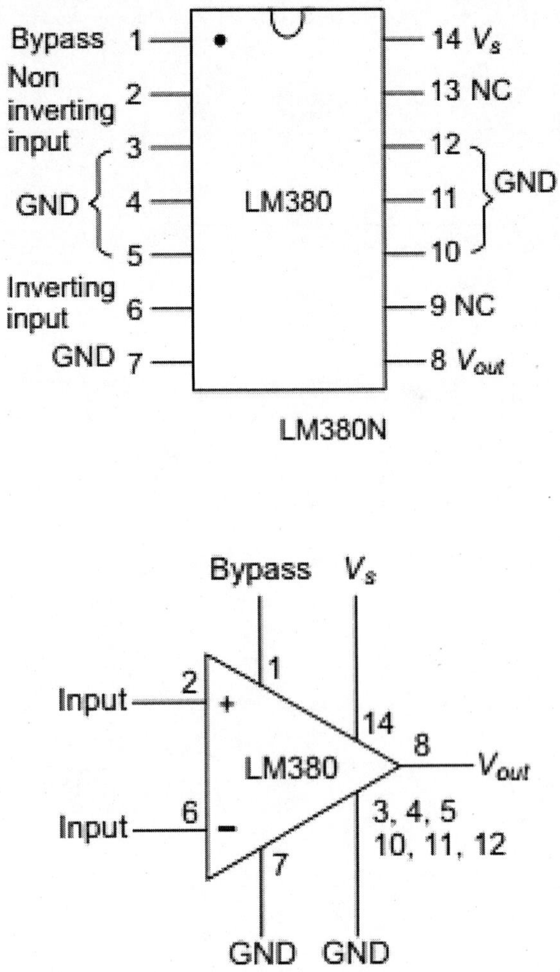

Fig.5.13.Pin diagram of LM380

LM380 circuit description:

It is connected of 4 stages,

i. PNP emitter follower

ii. Different amplifier

iii. Common emitter

iv. Emitter follower

(i) PNP Emitter follower:

The PNP Q_1-Q_2 differential pair is driven by the PNP Q_3-Q_4 input stage, which is an emitter follower made up of these transistors. Because Q_1 & Q_2 are PNP input transistors, the input can be referred to ground, allowing direct coupling to both the inverting and non-inverting terminals of the amplifier.

(ii) Differential Amplifier:

Q_7, R_3 and +V determine the current in the PNP differential pair Q_3-Q_4. The collector current of Q_9 is then established by the current mirror generated by transistors Q_7, Q_8, and accompanying resistors. Collector loads for the PNP differential pair are provided by transistors Q_5 and Q_6. The differential amplifier output is obtained at the junction of Q_4 and Q_6 transistors and used as an input to the common emitter voltage gain.

(iii) Common Emitter amplifier stage:

Transistor Q_9 forms the common emitter amplifier stage, with D_1, D_2, and Q_8 serving as current source loads. The capacitor C connected between the base and collector of Q_9 provides internal compensation and contributes to the upper cutoff frequency of 100 KHz. Because Q_7 and Q_8 form a current mirror, the current via D_1 and D_2 is about equal to the current through R_3. D_1 and D_2 are temperature compensating diodes for transistors Q_{10} and Q_{11}, as they have the same characteristics as Q_{11}'s base-emitter junctions. As a result, the current via Q_{10} and (Q_{11}-Q_{12}) is roughly equivalent to the current through diodes D_1 and D_2.

(iv) (Output stage) - Emitter follower:

The NPN transistors Q_{10} and Q_{11} combine to generate an emitter follower. The PNP transistor Q11 and the NPN transistor Q_{12} in conjunction create a device that has the power-handling capabilities of NPN transistors but the properties of a PNP transistor. The differential amplifier is brought into balance by the negative dc feedback that is applied by R_5, and as a result, the dc output voltage is maintained at +V/2. Connecting a bypass capacitor in the order of micro farads between the bypass terminal (pin 1) and ground (pin 7) is what needs to be done in order to disconnect the input stage from the supply voltage +V.The overall internal gain of the amplifier cannot be changed from its default setting of 50. On the other hand, benefit can be maximised by the utilisation of positive feedback.

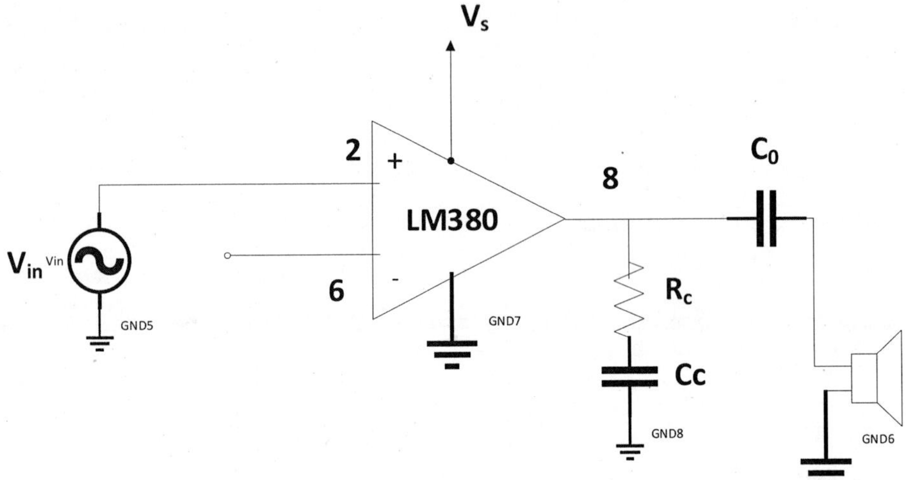

Audio Power Amplifier

Fig.5.14.Audio Power Amplifier

Figure 5.14 shown above is the connection of Audio power Amplifier. Amplifier requires very few external components because of the internal biasing, compensation & fixed gain.When the power amplifier is used in the non-inverting mode, the inverting terminal may either be shorted to ground or connected to ground through resistors and capacitors. This allows the amplifier to function in a configuration that does not flip the signal. When a power amplifier is used in the inverting mode, the non-inverting terminal can be either shorted to ground or returned to earth by a resistor or capacitor. This is also the case when the power amplifier is used in the normal mode. If the input has a high internal impedance, it is common practise to connect a capacitor between the inverting terminal and ground in electronic circuitry. To prevent oscillation between 5 and 10 MHz, a combination of resistors and capacitors (RC) should be connected to the output terminal (pin 8) as a preventative measure.C_C is the coupling capacitor that connects the output of the amplifier to the load, which is the 8-ohm loud speaker. Vin signal that is applied to the amplifier's non-inverting terminal will have its strength increased by the amplifier.

5.8 VIDEO AMPLIFIER

DESCRIPTION

Monolithic differential input and differential output characterise the 733, a wideband video amplifier that features differential input. It provides fixed gains of 10,100, or 400 without the need for any other components, and it also provides gain adjustments ranging from 10 all the way up to 400 with the use of an external resistor. Any gain choice can be utilised without the need for any external frequency adjustment components. By using the traditional series-shunt feedback from the emitter follower outputs to the inputs of the second stage, gain stability, a wide bandwidth, and low phase distortion can be achieved. The capability of the device to drive capacitive loads is made possible by the emitter follower outputs, which offer a low output impedance. The 733 is designed to function as a high-performance video and pulse amplifier in a variety of electronic systems, including communications, magnetic memory, display, and video recorders. The pin configuration and IC package of the video amplifier are displayed in Fig. 1, which can be found below.

FEATURES

- 120 MHz Bandwidth
- 250k£2 Input Resistance
- Selectable Gains of 10,100 and 400
- No Frequency Compensation Required

PIN CONFIGURATIONS

A & F PACKAGE

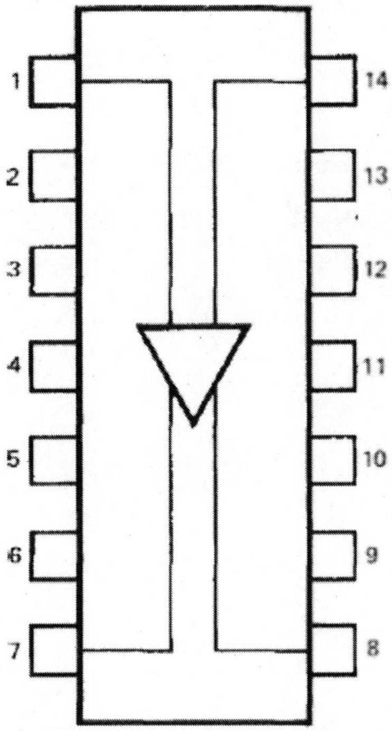

1. Input 2
2. NC
3. G_{2B} Gain Select
4. G_{1B} Gain Select
5. V^-
6. NC
7. Output 2
8. Output 1
9. NC
10. V^+
11. G_{1A} Gain Select
12. G_{2A} Gain Select
13. NC
14. Input 1

ORDER PART NOS.

μA733A/ μA733F/
μA733CA/ μA733CF

K PACKAGE

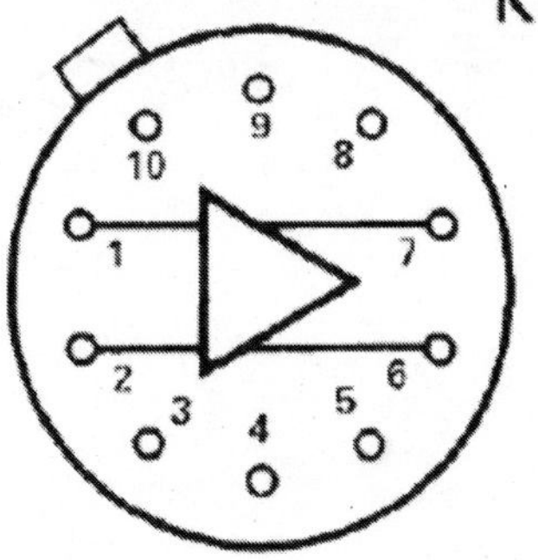

1. Input 1
2. Input 2
3. G_{2B} Gain Select
4. G_{1B} Gain Select
5. V^-
6. Output 2
7. Output 1
8. V^+
9. G_{1A} Gain Select
10. G_{2A} Gain Select

NOTE: Pin 5 connected to case.

ORDER PART NOS.
μA733K/μA733CK

Fig.5.15.Video Amplifier

5.9 ISOLATION AMPLIFIER

Isolation can be provided in a circuit by using either an isolation amplifier or a unity gain amplifier. Both of these types of amplifiers work in the same way. Therefore, there is no way for the power to be taken, used, or squandered within the circuit. The signal strength can be improved mostly thanks to the work of this amplifier. The output signal from the operational amplifier is identical to the input signal that was fed into the operational amplifier. Isolation and an electrical safety battier are both provided by these amplifiers for their respective uses. These amplifiers shield the patients from the electricity that is flowing away from them. They break the ohmic continuity of electrical signals between the input and output, and an isolated power source can be provided for both the input and the output of the device. Therefore, amplification of the low-level signals is possible.

An amplifier that does not have any conductive contact between its input and output parts is an example of an isolation amplifier. This type of amplifier is also known as an isolated amplifier. As a consequence of this, this amplifier provides ohmic isolation between the input/output terminals and the output/input terminals of the amplifier. This isolation needs to have a low quantity of leakage in addition to a significant amount of dielectric breakdown voltage. The normal numbers for the resistor and capacitor of an amplifier are as follows: the resistor should have 10 tera ohms, and the capacitor should have 10 picofarads. These values are found between the input and output terminals.When there is a significant difference in the common-mode voltage between the input and output sides, this type of amplifier is usually used. The ohmic circuitry is missing from the input ground all the way to the output ground in this amplifier.

FEATURES

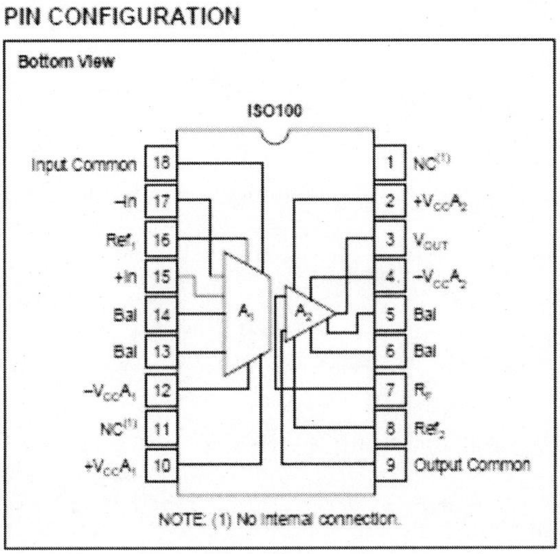

Fig.5.16.Pin configuration of Isolation Amplifier

1) Easy to use, similar to an OP AMP
2) VOUT/IIN = RF, Current Input
3) VOUT/VIN = RF/RIN, Voltage Input
4) 100% tested for breakdown:
5) 750V Continuous Isolation Voltage
6) Ultra-low leakage: 0.3mA, max, at
7) 240V/60Hz
8) Wide bandwidth: 60kHz
9) 18-PIN DIP PACKAGE

5.10 OPTO COUPLERS/OPTO ISOLATORS AND FIBRE OPTIC IC

Opto couplers, also known as opto isolators, are devices that contain both a light source and a light detector within the same housing. They are used to couple signals optically from one location to another, and they do this by ensuring that both points remain completely electrically isolated from one another. For the purpose of providing protection for the control circuit, an isolation of this kind is established between a low power control circuit and a high power output circuit.

Characteristics of opto coupler:

(i) Current Transfer Ratio:

It is defined as the ratio of output collector current (I_c) to the input forward current (I_f)

$$CTR = I_c/I_f * 100\%.$$

Its value depends on the devices used as source & detector.

(ii) Isolation voltage between input & output:

It is the highest voltage that can exist differentially between the input and the output without impacting the electrical isolation voltage. The electrical isolation voltage is defined in K Vrms with a relative humidity of 40 to 60%.

(iii) Response Time:

The response time of an optical coupler describes how quickly it can switch the state of its output. The detector transistor, the input current, and the load resistance all play a significant role in deter mining response time.

(iv) Common mode Rejection:

Even though the opto couplers are electrically isolated for dc & low frequency signals, an impulsive input signal (the signal which changes suddenly) can give rise to a displacement current $Ic= C_f*dv/dt$. This current can flow between input & output due to the capacitance C_f existing between input & output. This allows the noise to appear in the output. Depending on the type of light source & detector used we can get a variety of opto couplers. They are as follows,

LED – PHOTODIODE OPTO COUPLER

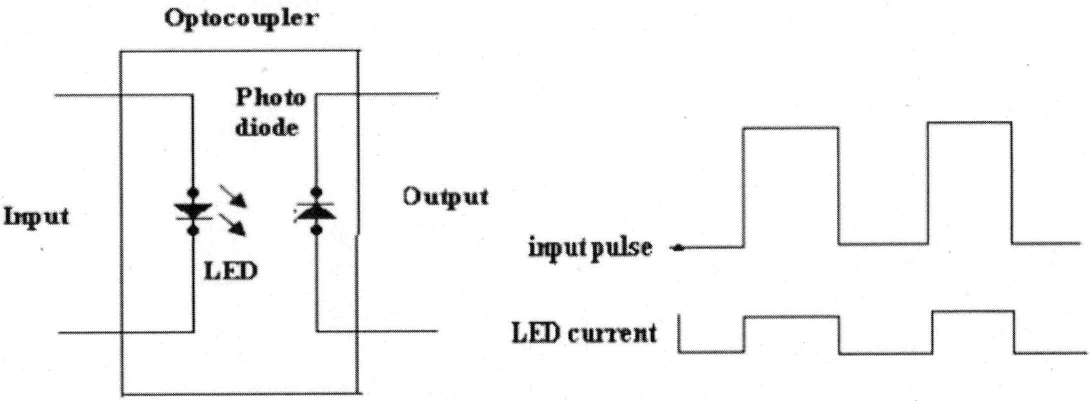

Fig.5.17. Opto coupler

Figure 5.17 depicts an LED photodiode and its waveforms, where the infrared LED serves as a light source and the photodiode serves as a detector. The photodiode has the benefit of being highly linear. When the input pulse goes high, the LED illuminates. It gives off light. This light is directed towards the photodiode. The photocurrent will begin to flow through the photodiode in response to this light. When the input pulse falls to zero, the LED goes off and the photocurrent via the photodiode falls to zero as well. As a result, the input pulse is connected to the output side.

Advantages of Opto coupler:

1. Control circuits are well protected due to electrical isolation.
2. Wideband signal transmission is possible.
3. Due to unidirectional signal transfer, noise from the output side does not get coupled to the input side.
4. Interfacing with logic circuits is easily possible.
5. It is small size & light weight device.

Disadvantages:

1. Slow speed.
2. Possibility of signal coupling for high power signals.

Applications:

Opto couplers are used basically to isolate low power circuits from high power circuits.At the same time the control signals are coupled from the control circuits to the high power circuits.

Some of such applications are,

i. AC to DC converters used for DC motor speed control

ii. High power choppers

iii. High power inverters

One of the most important applications of an opto coupler is to couple the base driving signals to a power transistor connected in a DC-DC chopper.

FIBRE OPTIC IC:

Optoelectronics and optoelectronic devices have advanced dramatically in recent years. Optics has the advantages of high bandwidth, parallelism, and reconfigurability. Electronic gadgets play an active role in information-handling systems. Thus, opto-electronic integrated circuits incorporate electrical and optical components, as well as optical interconnects. The optical connection medium is formed by the fibre. This type of connection media has a huge bandwidth, high data transmission speed, and immunity to mutual interference and cross-talk. Capacitive loading effects have no effect on them either. It reduces system size, reduces system power, and increases fan-out capabilities. The block diagram of an opto-electronic-integrated circuit is shown in Figure 5.18.It combines optical detection with electronic capabilities such as switching and amplification, as well as light transmission.

The main parts and major components of a two-link fiber-optic communication facility are depicted in Figure 5.18. An optical transmitter, such as an LED or a laser diode source with driver, is coupled to the fibre at the transmitting end. The LED or laser diode turns on and off in response to the bit stream. The signal from the first link is received by the repeaters. The signal is amplified, rearranged, and retimed before being sent to the next link. An avalanche photo-diode or PIN diode senses the attenuated and dispersed stream of light pulses in the repeater. The pulses are then processed before being transmitted to the second link. At the receiving end, the receiver converts incoming light pulses to electrical pulses. Electrical pulses are created and distributed.

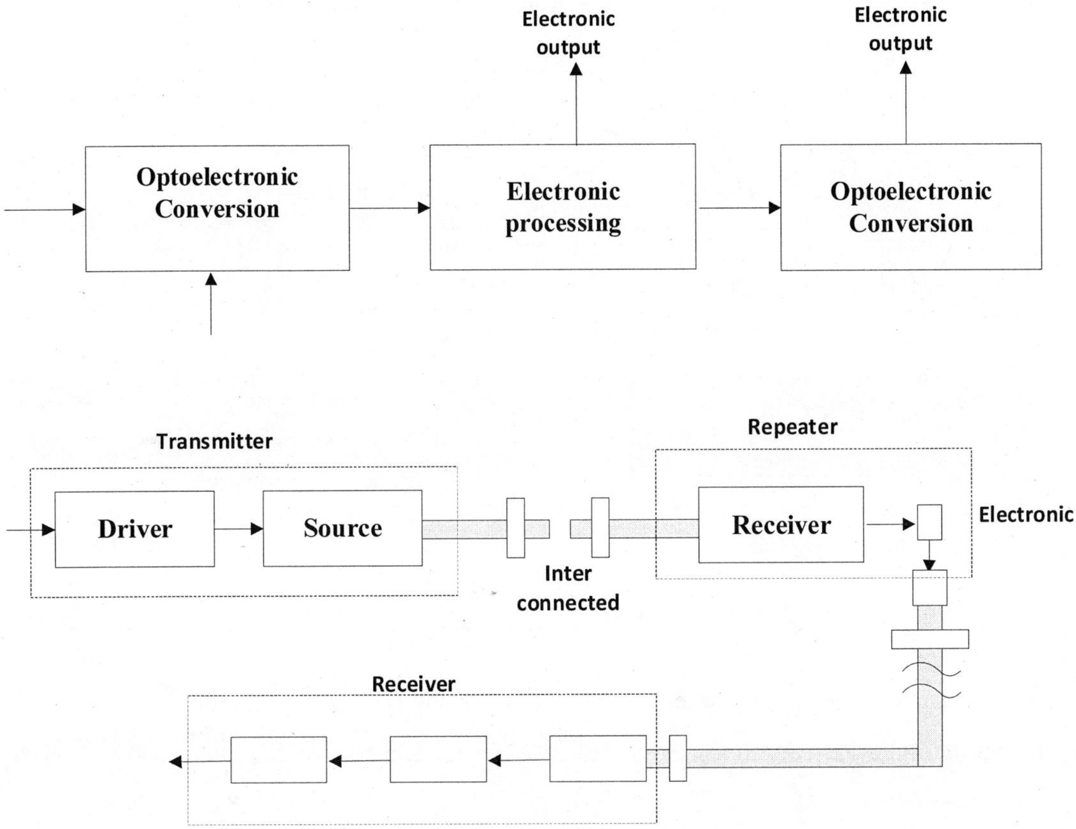

Fig.5.18. opto-electronic-integrated circuit

The total loss incurred by the fibre and all connections and splices throughout the length of the fibre optic communication line in a fibre-optic link must not exceed a particular minimal value. The lowest admissible received optical power P_r is determined by the detector type and the permitted error rate. Assume the optical source emits a total light flux of power P_t, as indicated in figure 5.14.4.Due to the inefficiency of linking the diode to the fibre end, only a portion of the light is delivered through the fibre. As a result, L_{pt} is lost in part.

Similarly, the connector at the opposite end of the fibre introduces a Lc insertion loss. Light travelling through the fibre experiences losses due to absorption and leakage, which can be expressed as Z_{Lf}, where Lf is the rate of loss and Z is the length of the fibre. L_s splice loss is also introduced for each of the Ns splices.

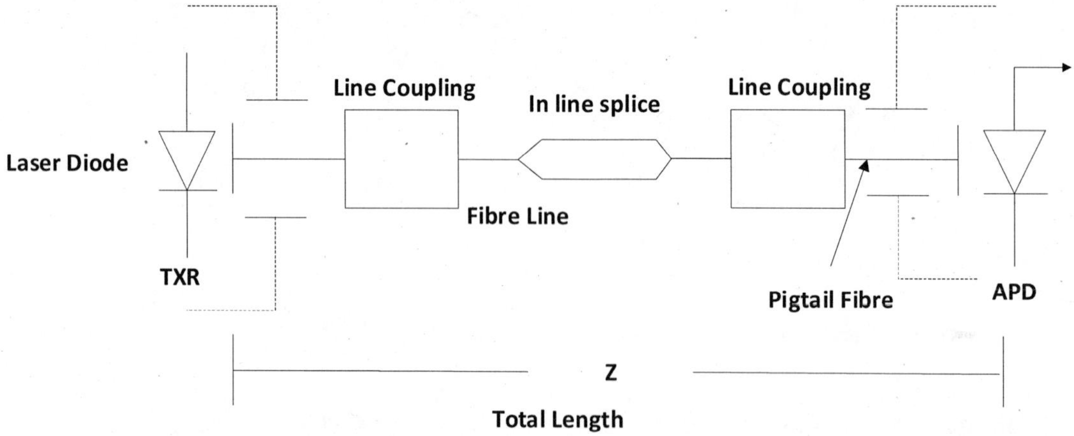

Fig.5.19.Opto Electronic Circuit

At the receiving end, the digital line connector shown feeds the light from the fibre in to the avalanche photo diode. This introduces a second part loss L_{pr} and a second connector loss L_c. The loss budget for the communication link is then, $P_t-P_r=M+L_{pt}+L_{pr}+N_cL_c+N_sL_s+Z_{Lf}$. Therefore, the maximum length of link Z is determined and limited by the losses.

5.11 Volatge Regulators

IC voltage regulators have the ability to produce a DC voltage output that is constant, regardless of the DC voltage that is being supplied. A voltage regulator is one of the electronic components that are used in the vast majority of everyday devices. It is essential for the efficient operation of a wide variety of digital electronic equipment to have a voltage that has been properly regulated and is free from fluctuations and noise levels. In the case of microcontrollers, for example, it is typically necessary to supply a voltage that has been smoothly regulated in order for the microcontroller to function normally.

If the input voltage is 230 V AC and you need a constant DC output voltage, you will need to utilize a step-down transformer in addition to rectifier and filter circuits. This is because you will need to maintain the output voltage at a constant level. The block diagrams are depicted in Figures 5.20 and 5.21 respectively.

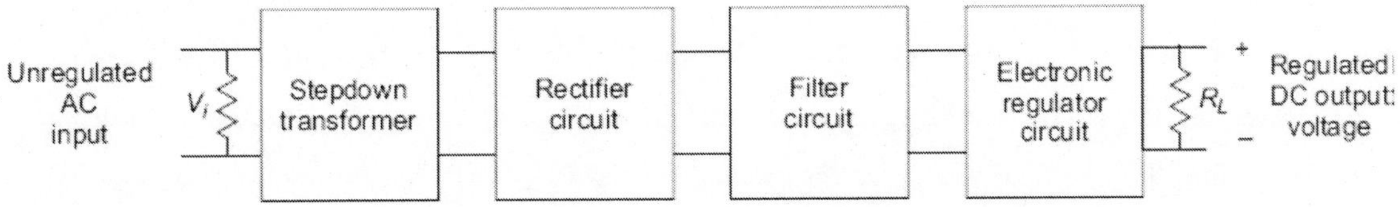

Fig. 5.20 Block Diagram of Voltage Regulator Circuit–AC Input

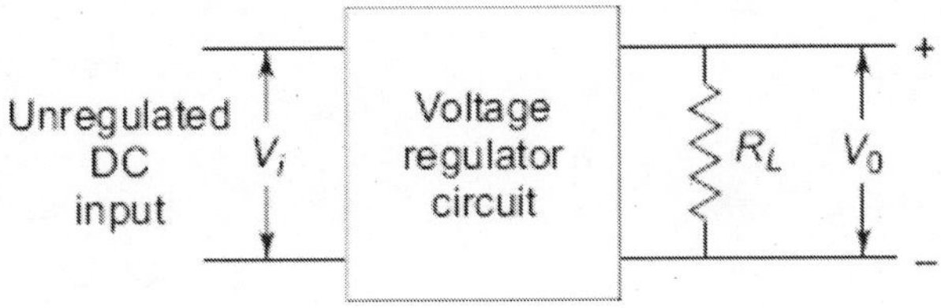

Fig. 5.21 Block Diagram of Voltage Regulator–DC Input

There is a classification of regulators as

1) Fixed voltage regulators (positive & negative)

2) Adjustable voltage regulators (positive & negative) and

3) Switching regulators.

The commonly used voltage regulator ICs are given below.

1. μA 723

2. LM 309

3. LM 105

4. CA 3085 A

5. 78 series: 7805, 7806, 7808, 7812, 7815—three terminal fixed positive voltage regulator ICs.

6. 79XX series: 7905, 7906, 7908, 7912, 7915—three terminal fixed negative voltage regulator ICs.

5.12 Three Terminal IC Voltage regulator

In accordance with the specifications of the power supply's design, a three-terminal IC regulator will be chosen. Unregulated DC voltage Vin is wired to the input terminals number one and three on the IC regulator.To compensate for voltage drops caused by the IC itself, the input voltage, or Vin, should be three volts higher than the DC output value that is intended.

The circuit on the inside will be comparable to one titled "series voltage regulator circuit using op amp."

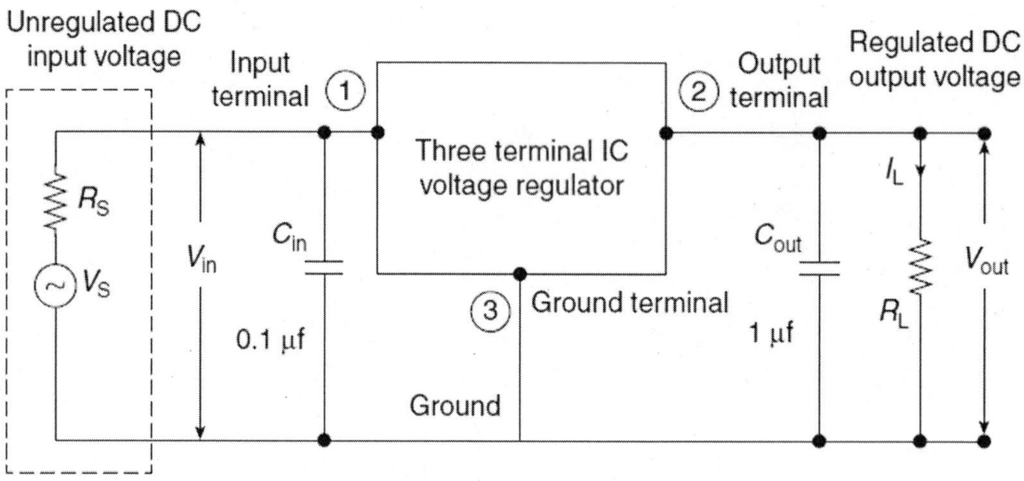

Fig.5.22 Three Terminal IC Voltage Regulator

The voltage of the regulated DC output will remain constant in accordance with the manufacturer's standards.

1. Load current variations generate rapid changes (transients) in output voltages. The output capacitor Cout in the circuit works to dampen these transitions and prevent them from occurring.
2. The size of IC voltage regulators has been significantly reduced, making them more compact.

Because there are just three pins on the integrated circuit, making connections for the circuit wiring is a simple process.

4. The IC regulator does not require any additional components from the outside world.

5. When there are fewer connections, the design of the circuit can be made more simply.

6. Exceptional performance while maintaining a high level of operational reliability.

7. Integrated circuits typically make use of heat sinks to reduce the amount of heating caused by greater currents.

8. Working with integrated circuits is reliable over a considerable amount of time.

5.12.1 Fixed Voltage Regulator Using IC LM 7805

The LM 7805 integrated circuit was developed to deliver a fixed DC voltage of 5 V. The terminals numbered 1, 2, and 3 are present on each of these regulator chips. The fundamental operating diagram can be shown in Figure 5.22.

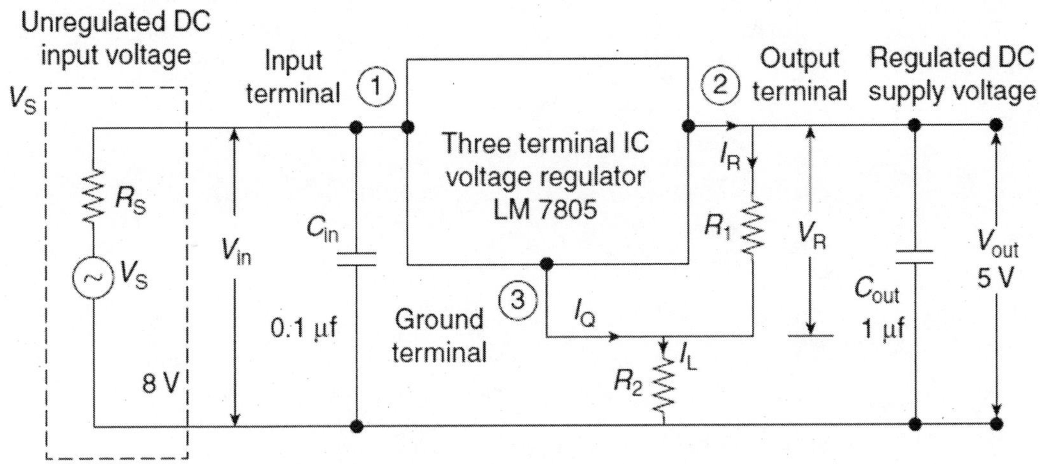

Fig.5.23 LM7805 based Positive Voltage regulator

1. The input terminal for uncontrolled DC input voltage is pin 1, often known as pin 1.

2. The output connector for the controlled DC output voltage is located at pin 2.

3. The common ground connection for the entire circuit is located at pin 3.

The input voltage is a DC voltage that has not been controlled. The input voltage receives extra filtering from the capacitance provided by C_{in}. Within the integrated circuit is a series voltage regulator circuit that makes use of an op amp. The changes in load current might create transients in the output voltage, which are suppressed by the output capacitor C_{out}.

5.12.2 Fixed Voltage Regulator Using IC LM 7905

A voltage with no regulation is placed across the terminals (1) and (3) of the component. A negative output voltage that has been regulated will be present across terminals (2) and (3).When the voltage regulator IC 7905 is employed, the output voltage is a negative value.

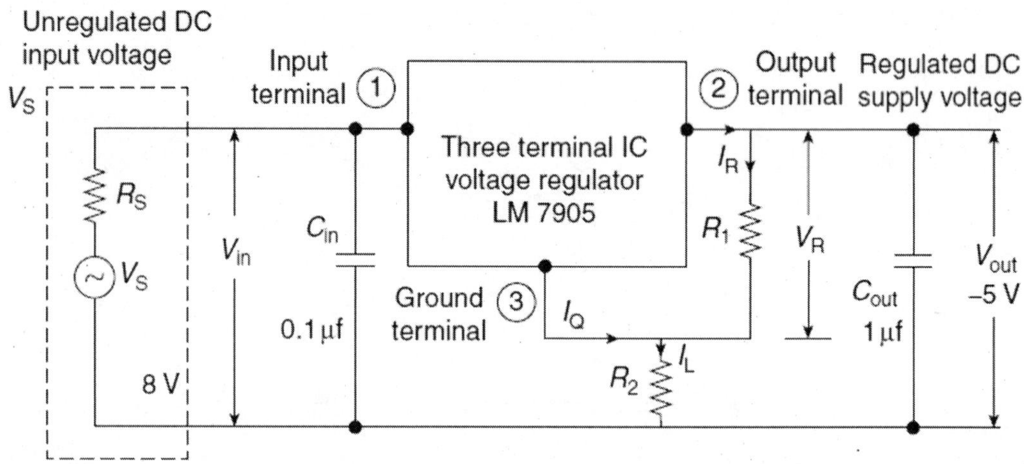

Fig.5.24 LM7905 based Negative Voltage regulator

5.13 IC 723 Voltage Regulator

Figure 5.25 is a functional block diagram of the A 723 (14-pin DIP) high voltage regulator. This regulator comes in a dual inline package. Zener diode D1 is the component that makes up IC 723's voltage regulator. Zener voltage functions as reference voltage V_{ref} (7.15 V). Pin 6 is the point of connection for it. Op amp functions as a 'error amplifier' in a circuit. The output terminal is connected to the base terminal of the control transistor T1 and the collector terminal of the current limiting (sensing) transistor T_2 via an internal connection. + V_{CC} is linked to pin 12 on the outside of the IC, and -V_{CC} is connected to pin 7 on the inside of the IC. Pin 10 is connected to the controlled output voltage of the pass transistor, hence this transistor functions as an emitter follower. It has an external connection to the inverting input terminal of the error amplifier (pin 4), which is denoted by the notation "INV." It does this by comparing a sample of the output voltage that is applied to the inverting input terminal of the operational amplifier with the reference voltage V_{ref} that is connected externally to the non-inverting input terminal of the operational amplifier.

The error signal is the voltage that is obtained from the output of the operational amplifier. It is connected to the base terminal of the series pass transistor T_1, which is the current source. Fluctuations in the magnitudes of error inputs to the error amplifier lead to fluctuations in the conduction of pass transistor T1, also known as a variation in base current. As a result, the conductivity of transistor T_1 is determined by the erroneous signal. Variations in the conduction of the transistor (T_1) create variations in the magnitudes of the voltage corrected element (V_{CE}), which in turn corrects for variations in the output voltages of the regulator IC 723. Because of this, the voltage of the load is automatically adjusted in order to keep the output voltage of the 723 IC voltage regulators constant.

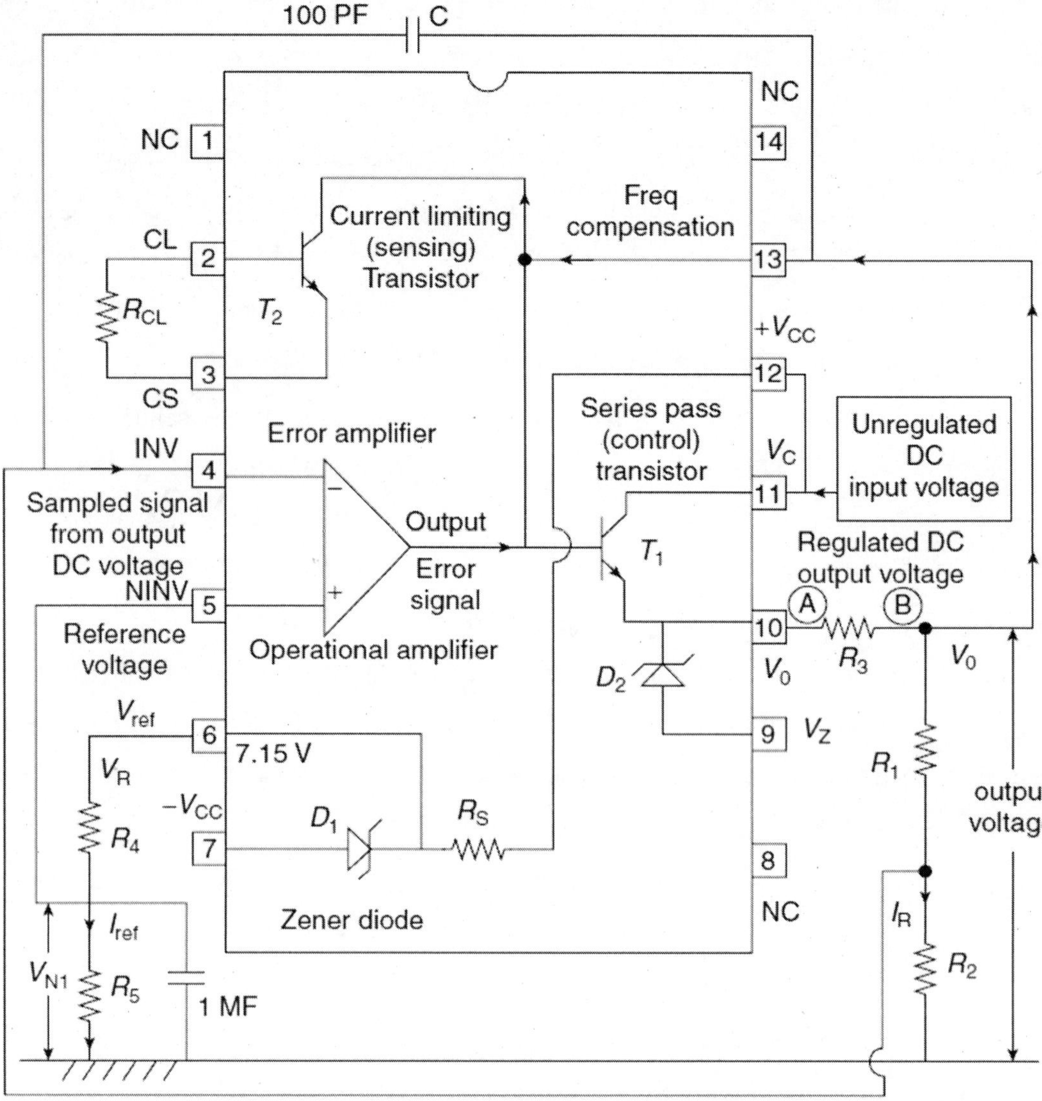

Fig.5.25 Diagram of 723 IC

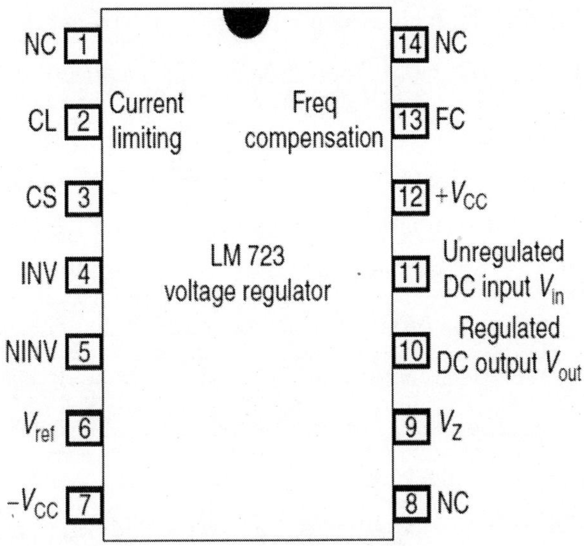

Fig.5.26 Pin Diagram of LM723 Voltage regulator

5.13.1 Working principle

1. A direct current voltage that has not been regulated is supplied to the collector terminal of the series pass (control) transistor T_1 (pin 11).

2. A representation of the output voltage is transferred from pin 10 (the output) to the inverting input terminal (INV) of the error amplifier (the operational amplifier) using pin 4.

3. A portion of the reference voltage V_{ref} (7.15 V) is connected to the non-inverting input terminal (NINV) (pin 5) of the error amplifier. This is accomplished by utilizing potential divider R_4 and R_5 to create a voltage divider across the zener diode (pin 6).

4. The error amplifier's output voltage is wired to the input terminal (base terminal) of the series pass transistor T_1, which is connected across terminals 10 and 11.

5. $V_{NINV} = V_{ref}\frac{R_5}{R_5+R_4}$

6. The pass transistor performs the function of an emitter follower, and its output is coupled to the external load across which the output voltage is created in order to supply voltage. Current through R_1 and R_2 is given by, $I_R = \frac{V_0-V_{ref}}{R_1}, R_2 = \frac{V_{ref}}{I_R}, R_3 = R_1$

If the output voltage 'V_0' across the load goes up as a result of any changes in the DC voltage of the output, then the voltage at the INV input of the error amplifier will go up as well. At its input terminals, the error amplifier makes a comparison between a sample of the voltage at the output and the reference voltage known as V_{ref}. The difference between the two voltages is that the reduced effective input signal of the error amplifier is caused by the increased reference voltage at the non-inverting terminal and the increased feedback signal at the inverting terminal. Error amplifier's output voltage drops as a result of the problem. It is connected to the base terminal of the series pass transistor T1, which operates as an emitter follower, so the input signal can reach it. A lower error signal will result in a lower forward bias being applied to transistor T_1, which will, in turn, cause a drop in the load current I_L being produced by that transistor. As a result, the voltage across the load decreases in order to keep the output voltage at a constant level.

When the output voltage 'V_0'across the load drops, a similar rationale might be made. If the output voltage goes down, the effective input signal of the error amplifier will go up. This, in turn, will cause the error input signal that is given to the pass transistor 'T_1' to go up as well. As a result, there is an increase in the forward-bias to the pass transistor. Because of this, there is a rise in the load current that is flowing through the load. An increase in load current results in a rise in output voltage relative to the voltage that was supposed to be constant.

UNIT SUMMARY

- In electronic circuits, operational amplifiers, or op-amps, are frequently used to produce a variety of waveforms that are then put to use in analogue and digital devices for purposes including timing, control, and other activities.
- Using a square-wave generator that is coupled to an integrator allows one to create a triangle waveform generator.
- The rise time of the sawtooth wave is several times longer than the fall time, or the fall time is significantly longer in comparison to the rise period. Sawtooth waveform can be generated using a triangular wave generator with just minor adjustments.

- ICL8038 is capable of frequency modulation in some applications.
- Signetics Corporation was the company that initially presented the Type SE555/NE555 integrated circuit timer to the market.
- The 555 timer can function with a direct current (dc) source voltage ranging from +5V to +18V. In addition to being exceedingly adaptable, it is compatible with TTL and CMOS logic circuits as well as op-amp circuits.
- Oscillator, pulse generator, square and ramp wave generator, one-shot multivibrator, safety alarm and timer circuits, and traffic light controllers are some of the applications of the 555 timer. The time delay that can be provided by the IC can range anywhere from microseconds to hours.
- The following special purpose integrated circuits are discussed: voltage-to-frequency converters and frequency-to-voltage converters, opto-couplers, isolation amplifiers, fiber-optic ICs, and companders.
- A Voltage-to-Frequency (V/F) converter generates an output signal whose frequency at any given instant is a function of the external control input voltage. This relationship is represented by the equation $f_o = k_v V_i$, where k_v is the sensitivity of the V/F converter expressed in Hz/V.
- An F/V converter, also known as a frequency-to-voltage converter, generates an output voltage whose amplitude is a function of the frequency of the input signal and is represented by the equation $V_o = k_f f_i$.
- Opto-couplers find their most common use in computer terminals, industrial system controllers, measuring instruments, and signal transmission systems with varying operating voltages and characteristic impedances. Other uses include controlling industrial systems.
- The primary benefits of opto-couplers include electrical isolation of several thousand volts, a very small response time that enables its use in data applications operating in the megahertz range of pulse frequencies, the ability to transmit wideband signals, easy interfacing capability, portability, better efficiency than isolation transformers and relays,

and the elimination of problems caused by ambient conditions such as noise, transients, and contact bounce, among other things.

- The isolation amplifier ensures that there is no physical contact established between the input and the output. In applications that require a very large common-mode voltage difference on the order of several thousand volts between the input and output portions, it is utilised. Linearity is another requirement for these applications. The formation of this isolation barrier is made possible by an optical coupler that functions as a current translator on a 1:1 scale. An input amplifier, a GaAs Light-Emitting Diode (LED), a silicon photo diode, and an output amplifier make up the majority of the isolation amplifiers. These isolation amplifiers are hybrid integrated circuits.
- Because of the lower cost of using ICs to manufacture circuits, IC regulators are quite popular. This is one reason for their popularity. Integrated circuit regulator circuits offer a design that is both very reliable and adaptable. Miniaturisation of power supply has allowed for advanced communication via the iPad, cell phones with 4G technology and 4S (Samsung), and HTC smart phones. This has been made possible by the extremely low amount of power that is consumed by these devices.

Multiple Choice Questions

1. The following circuit is a part of 555 IC internal circuits
   (a) Peak detector
   (b) Two comparators
   (c) Comparator
   (d) Voltage amplifier

[Ans. (b)]

2. The Frequency of oscillation of squarewave generator is
   (a) $f0 = \frac{1}{2RC\ln\left(\frac{2R1+R2}{R2}\right)}$
   (b) $f0 = \frac{1}{4RC\ln\left(\frac{2R1+R2}{R2}\right)}$

(c) $f0 = \frac{1}{2\pi RC}$

(d) $f0 = \frac{1}{2RC}$

[Ans. (a)]

3. Zener diodes are used to get

(a) Perfect Square waveform

(b) Perfect Trainagular wave

(c) Sine wave

(d) No waveform

[Ans.(a)]

4. Sinewave generators is also called as

(a) Oscillator

(b) Amplifier

(c) Rectifier

(d) chopper

[Ans.(a)]

5. Input to Voltage to frequency converter

(a) Voltage

(b) Current

(c) Frequency

(d) Power

[Ans.(a)]

5. Three-terminal IC voltage regulator

(a) 723 (b) 7805

(c) 741 (d) 555

[Ans. (b)]

6. A regulated power supply circuit consists of the following components

   (a) Operational amplifier as error amplifier

   (b) Feedback network

   (c) Series pass Transistor

   (d) All of them

[Ans. (d)]

Answers of Multiple Choice Questions

Short and Long Answer Type Questions

1. Explain the Pin diagram of 555 IC.
2. Discuss the working of Isolation Amplifier.
3. Explain the working of Opto-coupler.
4. Explain the operation of square-wave generator using op-amp with capacitor and output voltage waveforms. How can you obtain a non-symmetrical square-wave?
5. Draw the circuit of an astable multivibrator using op-amp and derive the expression for its frequency of oscillations. How will you modify this circuit to have independent control of ON and OFF time durations?
6. Draw the internal block diagram of IC 723 voltage regulator and explain the function of each block.

REFERENCES AND SUGGESTED READINGS

1. Coughlin, R. F. and Frederick F. Driscoll (2001). Operational Amplifiers and Linear Integrated Circuits, Sixth Edition. Upper Saddle River, NJ: Prentice Hall.
2. Dailey, Denton J. (1989) Operational Amplifiers and Linear Integrated Circuits: Theory and Applications. New York: The McGraw-Hill Companies.
3. Franco, Sergio (2002). Design with Operational Amplifiers and Analog Integrated Circuits. New York: The McGraw-Hill Companies.

4. Gayakwad, Ramakant A. (1999). Op-Amps and Linear Integrated Circuits. Fourth edition. Upper Saddle River, NJ: Prentice Hall Inc.
5. Jacob, J. Michael (1996) Applications and Design with Analog Integrated Circuits. Upper Saddle River, NJ: Prentice Hall Inc.
6. *Ramakanth A. Gayakwad* (2000),Op-Amps and Linear Integrated Circuits -*, Prentice-Hall, 4th edition.*
7. S.Salivahanan,V S Kanchana Bhaskaran(2015),Linear Integrated Circuits, Second Edition, Tata McGrawa Hill Edition.
8. B.Visweswara Rao(2015), Linear Integrated Circuits, First Edition, Pearson India Education Services Pvt. Ltd

Dynamic QR Code for Further Reading

REFERENCES FOR FURTHER LEARNING

1. Coughlin, R. F. and Frederick F. Driscoll (2001). Operational Amplifiers and Linear Integrated Circuits, Sixth Edition. Upper Saddle River, NJ: Prentice Hall.
2. Dailey, Denton J. (1989) Operational Amplifiers and Linear Integrated Circuits: Theory and Applications. New York: The McGraw-Hill Companies.
3. Franco, Sergio (2002). Design with Operational Amplifiers and Analog Integrated Circuits.New York: The McGraw-Hill Companies.
4. Gayakwad, Ramakant A. (1999). Op-Amps and Linear Integrated Circuits. Fourth edition.Upper Saddle River, NJ: Prentice Hall Inc.
5. Jacob, J. Michael (1996) Applications and Design with Analog Integrated Circuits. Upper Saddle River, NJ: Prentice Hall Inc.
6. *Ramakanth A. Gayakwad* (2000),Op-Amps and Linear Integrated Circuits -, *Prentice-Hall, 4th edition.*
7. S.Salivahanan,V S Kanchana Bhaskaran(2015),Linear Integrated Circuits, Second Edition, Tata McGrawa Hill Edition.
8. B.Visweswara Rao(2015), Linear Integrated Circuits, First Edition, Pearson India Education Services Pvt. Ltd.

CO AND PO ATTAINMENT TABLE

Course outcomes (COs) for this course can be mapped with the programme outcomes (POs) after the completion of the course and a correlation can be made for the attainment of POs to analyze the gap. After proper analysis of the gap in the attainment of POs necessary measures can be taken to overcome the gaps.

Table for co-po attainment

Course Outcomes	1-Weak Correlation, 2-Medium Correlation, 3-Strong Correlation						
CO/PO	**PO-1**	**PO-2**	**PO-3**	**PO-4**	**PO-5**	**PO-6**	**PO-7**
CO-1							
CO-2							
CO-3							
CO-4							
CO-5							

INDEX